PROBLEM-BASED LEARNING IN THE LIFE SCIENCE CLASSROOM

K–12

TOM J. MCCONNELL · JOYCE PARKER · JANET EBERHARDT

Arlington, Virginia

Claire Reinburg, Director
Wendy Rubin, Managing Editor
Rachel Ledbetter, Associate Editor
Amanda O'Brien, Associate Editor
Donna Yudkin, Book Acquisitions Coordinator

Art and Design
Will Thomas Jr., Director
Himabindu Bichali, Graphic Designer, cover and interior design

Printing and Production
Catherine Lorrain, Director

National Science Teachers Association
David L. Evans, Executive Director
David Beacom, Publisher

1840 Wilson Blvd., Arlington, VA 22201
www.nsta.org/store
For customer service inquiries, please call 800-277-5300.

21 20 19 18 5 4 3 2

NSTA is committed to publishing material that promotes the best in inquiry-based science education. However, conditions of actual use may vary, and the safety procedures and practices described in this book are intended to serve only as a guide. Additional precautionary measures may be required. NSTA and the authors do not warrant or represent that the procedures and practices in this book meet any safety code or standard of federal, state, or local regulations. NSTA and the authors disclaim any liability for personal injury or damage to property arising out of or relating to the use of this book, including any of the recommendations, instructions, or materials contained therein.

Library of Congress Cataloging-in-Publication Data
Names: McConnell, Tom J., 1962 , author. | Parker, Joyce, 1955- , author. | Eberhardt, Janet, 1952- , author.
Title: Problem-based learning in the life science classroom, K-12 / Tom J. McConnell, Joyce Parker, Janet Eberhardt.
Description: Arlington. VA : National Science Teachers Association, 2016. | Includes index.
Identifiers: LCCN 2016013523 (print) | LCCN 2016024967 (ebook) | ISBN 9781941316207 (print) | ISBN 9781941316696 (e-book)
Subjects: LCSH: Biology--Study and teaching (Elementary) | Biology--Study and teaching (Secondary) | Problem-based learning.
Classification: LCC QH315 .M44 2016 (print) | LCC QH315 (ebook) | DDC 570.71--dc23
LC record available at *https://lccn.loc.gov/2016013523*

CONTENTS

PREFACE

In science education, there are numerous strategies designed to promote learners' ability to apply science understanding to authentic situations and build connections between concepts (Bybee, Powell, and Trowbridge 2008). Problem-based learning (PBL; Delisle 1997; Gijbels et al. 2005; Torp and Sage 2002) is one of these strategies. PBL originated as a teaching model in medical schools (Barrows 1986; Schmidt 1983) and is relevant for a wide variety of subjects. Science education, in particular, lends itself to the PBL structure because of the many authentic problems that reflect concepts included in state science standards and the *Next Generation Science Standards* (*NGSS*; NGSS Lead States 2013).

The Problem-Based Learning Framework

PBL is a teaching strategy built on a constructivist epistemology (Savery and Duffy 1995) that presents learners with authentic and rich, but incompletely defined, scenarios. These "problems" represent science as it appears in the real world, giving learners a reason to collaborate with others to analyze the problem, ask questions, pose hypotheses, identify information needed to solve the problem, and find information through literature searches and scientific investigations. The analysis process leads learners to co-construct a proposed solution (Torp and Sage 2002).

One of the strengths of the PBL framework is that learners are active drivers of the learning process and can develop a deeper understanding of the concepts related to the problem starting from many different levels of prior understanding. PBL is an effective strategy for both novices and advanced learners. PBL is also flexible enough to be useful in nearly any science context.

One of the challenges for teachers and educational planners, though, is that implementing PBL for the classroom requires advance planning. An effective problem should be authentic, and the challenges presented in the problems need to be both structured and ill-defined to allow genuine and productive exploration by students. Dan Meyer (2010) suggested that these problems help students learn to be "patient problem solvers. " For most instructors, getting started with PBL in the science classroom is easiest with existing problems. However, there are very few tested PBL problems available in print or on the internet. Valuable resources exist that describe in general what PBL is, how to develop lessons, and how PBL can help students, but curriculum resources are much harder to find.

In this book, we present a discussion of the PBL structure and its application for the K–12 science classroom. We also share a collection of PBL problems developed as part of the PBL Project for Teachers, a National Science Foundation–funded professional development program that used the PBL framework to help teachers develop a deeper understanding of science concepts in eight different content strands (McConnell et al. 2008; McConnell, Parker, and Eberhardt 2013). Each content strand had a group of participants and facilitators who focused on specific concepts within one of the science disciplines, such as genetics, weather, or force and motion. The problems presented in this book were developed by content experts who facilitated the workshops and revised the problems over the course of four iterations of the workshops. Through our work to test and revise the problems, we have developed a structure for the written problem that we feel will help educators implement the plans in classrooms.

Because the problems have been tested with teachers, we have published research describing the effectiveness of the problems in influencing teachers' science content knowledge (McConnell, Parker, and Eberhardt 2013). The research revealed that individuals with very little familiarity with science concepts can learn new ideas using the PBL structure and that the same problem can also help experienced science learners with a high degree of prior knowledge to refine their understanding and learn to better explain the mechanisms for scientific phenomena.

Alignment With the *Next Generation Science Standards*

To ensure that the problems presented here are useful to science teachers, we have included information aligning the objectives and learning outcomes for each problem with the *NGSS*. The *NGSS* present performance expectations for science education that describe three intertwined dimensions of science learning: science and engineering practices (SEPs), disciplinary core ideas (DCIs), and crosscutting concepts (CCs). The *NGSS* emphasize learning outcomes in which students integrate the SEPs, DCIs, and CCs in a seamless way, resulting in flexible and widely applicable understanding.

The learning targets for the PBL problems included in this book were originally written with attention to the science concepts—what the *NGSS* calls DCIs. The aim of the PBL Project was to enhance teachers' knowledge of these core ideas. But implicit in the design of the PBL process is the need for learners to use the practices of science and make connections between concepts that reflect the CCs listed in the standards. PBL problems align well with the *NGSS* because these real-world situations present problems in a similar framework: SEPs, DCIs, and CCs are natural parts of the problems. We describe the alignment of the PBL problems with the *NGSS* in more detail in Chapter 2. As states begin to adopt or adapt these standards into state standards, Chapter 2 should help teachers and teacher educators fit the problems into their local curricula.

Intended Audiences and Organization of the Book

As mentioned earlier, the PBL problems in this book have been shown to be effective learning tools for learners with differing levels of prior knowledge. Some of the teachers who participated in the PBL Project used problems from the workshops in their K–12 classrooms, and facilitators with the project have also incorporated problems from this collection into university courses.

Chapter 2 discusses the alignment of the PBL problems and analytical framework with the *NGSS*. Chapter 3 describes strategies for facilitating the PBL lessons. In Chapter 4, we share tips for the classroom teacher on combining PBL with other activities in your curriculum, grouping students, managing information, and assessing student learning through the PBL process.

Chapters 5–8 present the problems we have designed and tested in four content strands: elementary life cycles, ecology, genetics, and cellular metabolism. For each problem, we include a table outlining alignment with the *NGSS*; interdisciplinary connections; resources and/or investigations that help provide relevant information about the science concept and problem; a teacher guide with the problem context, a model response to the challenge question(s) about the problem, and (in some problems) an activity guide; and assessment questions we used to evaluate learning, with model responses. To help you locate the problems that are most appropriate for your classroom, we have included a catalog of problems (pp. xiii–xiv); the catalog is in tabular format and will let you scan the list of problems by the topics, keywords and concepts, and grade bands for which the problems were written.

We hope that this collection of problems will serve as a model for educators who want to design and develop problems of their own. Some of the problems in this book relate to local ecosystems and examples that reflect contexts relevant to Michigan, where the PBL Project was located. A teacher in a place that does not share similar conditions may find that his or her students cannot relate to the scenario described in the problem. In those cases, we encourage teachers to modify and adapt problems to fit contexts familiar to their own students. Chapter 9 discusses how teachers can modify the problems in this book or design their own problems for PBL lessons.

This book is intended as the first volume in a PBL series. We present life science problems in this volume, and we will offer problems specifically written for teaching Earth-space science and physics in future volumes. A fourth volume will contain tips and examples for planners of teacher professional development programs. As you modify and implement lessons from these books, you can begin to develop your own problems that meet the needs of your students.

Safe and Ethical Practices in the Science Classroom

With hands-on, process- and inquiry-based laboratory or field activities, the teaching and learning of science today can be both effective and exciting. Successful science teaching needs to address potential safety issues. Throughout this book, safety precautions are provided for investigations and need to be adopted and enforced to provide for a safer learning and teaching experience.

Additional applicable standard operating procedures can be found in the National Science Teachers Association's (NSTA's) Safety in the Science Classroom, Laboratory, or Field Sites document (*www.nsta.org/docs/SafetyInTheScienceClassroomLabAndField.pdf*).

Science teaching needs to deal with animals in a safe and ethical way. We encourage teachers to review the NSTA position statement Responsible Use of Live Animals and Dissection in the Science Classroom (*www.nsta.org/about/positions/animals.aspx*). For information on field trip safety, read the NSTA Safety Advisory Board paper called "Field Trip Safety" (*www.nsta.org/docs/FieldTripSafety.pdf*).

Please note that the safety precautions of each activity are based, in part, on use of the recommended materials and instructions, legal safety standards, and better professional practices. Selection of alternative materials or procedures for these activities may jeopardize the level of safety and therefore is at the user's own risk.

References

Barrows, H. S. 1986. A taxonomy of problem-based learning methods. *Medical Education* 20 (6): 481–486.

Bybee, R. W., J. C. Powell, and L. W. Trowbridge. 2008. *Teaching secondary school science: Strategies for developing scientific literacy*. Upper Saddle River, NJ: Prentice Hall.

Delisle, R. 1997. *How to use problem-based learning in the classroom*. Alexandria, VA: Association for Supervision and Curriculum Development.

Gijbels, D., F. Dochy, P. Van den Bossche, and M. Segers. 2005. Effects of problem-based learning: A meta-analysis from the angle of assessment. *Review of Educational Research* 75 (1): 27–61.

McConnell, T. J., J. Eberhardt, J. M. Parker, J. C. Stanaway, M. A. Lundeberg, M. J. Koehler, M. Urban-Lurain, and PBL Project staff. 2008. The PBL Project for Teachers: Using problem-based learning to guide K–12 science teachers' professional learning. *MSTA Journal* 53 (1): 16–21.

McConnell, T. J., J. M. Parker, and J. Eberhardt. 2013. Problem-based learning as an effective strategy for science teacher professional development. *The Clearing House: A Journal of Educational Strategies, Issues and Ideas* 86 (6): 216–223.

Meyer, D. 2010. TED: Math class needs a makeover [video]. *www.ted.com/talks/dan_meyer_math_curriculum_makeover?language=en.*

NGSS Lead States. 2013. *Next Generation Science Standards: For states, by states*. Washington, DC: National Academies Press. *www.nextgenscience.org/next-generation-science-standards.*

Savery, J. R., and T. M. Duffy. 1995. Problem based learning: An instructional model and its constructivist framework. *Educational Technology* 35 (5): 31–38.

Schmidt, H. G. 1983. Problem-based learning: Rationale and description. *Medical Education* 17 (1): 11–16.

Torp, L., and S. Sage. 2002. *Problems as possibilities: Problem-based learning for K–16 Education.* 2nd ed. Alexandria, VA: Association for Supervision and Curriculum Development.

CATALOG OF PROBLEMS

Problem	Page Number	Keywords and Concepts	Grade Band			
			Grades K–2	Grades 3–5	Grades 6–8	Grades 9–12
CHAPTER 5: ELEMENTARY LIFE CYCLES PROBLEMS						
1. Baby Hamster	68	Mammalian life cycle, needs of living things	•	•		
2. Wogs and Wasps	75	Insect and amphibian life cycles, needs of living things	•	•	•	
3. Humongous Fungus	83	Fungus life cycle, needs of living things		•	•	
4. Baby, Baby Pear	89	Plant life cycles, needs of living things	•	•		
CHAPTER 6: ECOLOGY PROBLEMS						
1. Where's Percho?	106	Competition, food webs		•	•	•
2. Lake Michigan—A Fragile Ecosystem	114	Competition, food webs, invasive species			•	•
3. Bottom Dwellers	121	Eutrophication, water quality indicators, water pollution, human impacts on the environment			•	•
4. Bogged Down	132	Ecological succession, plant dispersal, competition in plants			•	•
5. The Purple Menace	138	Competition, invasive species		•	•	•
CHAPTER 7: GENETICS PROBLEMS						
1. Pale Cats	154	Genotype to phenotype, inheritance of traits			•	•
2. Black and White and Spots All Over	162	Codominance, inheritance of traits			•	•
3. Calico Cats	169	Sex-linked and multigenic traits			•	•
4. Cat Puzzles	176	Patterns of inheritance, genotype to phenotype, codominance, sex-linked and multigenic traits			•	•

Problem	Page Number	Keywords and Concepts	Grade Band			
			Grades K–2	Grades 3–5	Grades 6–8	Grades 9–12
CHAPTER 8: CELLULAR METABOLISM PROBLEMS						
1. Torching Marshmallows	189	Cellular respiration		•	•	•
2. Mysterious Mass	202	Photosynthesis		•	•	•
3. Why We Are Not What We Eat	212	Digestion, biosynthesis			•	•

ACKNOWLEDGMENTS

We wish to thank the following individuals who helped design, revise, and facilitate the problem-based learning (PBL) lessons presented in this book. Their expertise and insight were instrumental in the development of the problems and tips on facilitating PBL learning.

Michigan State University

Merle Heidemann

Kazuya Fujita

Susan Jackson

Christopher Reznich

Mary Jane Rice

Lansing Community College

Alex Azima

K. Rodney Cranson

Christel Marschall

Dennis McGroarty

Teresa Schultz

Jeannine Stanaway

Mott Community College

Judith Ruddock

DeWitt High School

Mark Servis

Retired Faculty

Roberta Jacobowitz (Otto Middle School)

Barbara Neureither (Holt High School)

Zandy Zweering (Williamston Middle School)

Ingham Intermediate School District

Theron Blakeslee

Martha Couretas

ABOUT THE AUTHORS

Tom J. McConnell is an associate professor of science education in the Department of Biology at Ball State University, Muncie, Indiana. He teaches science teaching methods courses for elementary and secondary education majors and graduate students, as well as a biology content course for elementary teachers. His research focuses on the influence of professional development on teacher learning and student achievement, and it focuses on curriculum development for teacher education programs. He is also an active member of the Hoosier Association of Science Teachers and the National Association of Research in Science Teaching.

Joyce Parker is an assistant professor in the Department of Geological Sciences at Michigan State University. She teaches a capstone course for prospective secondary science teachers. Her research focuses on student understanding of environmental issues. She is an active member of the National Association of Research Science Teaching and the Michigan Science Teachers Association.

Janet Eberhardt is an emeritus teacher educator and former assistant director of the Division of Science and Mathematics Education at Michigan State University. She has served as a consultant with the Great Lakes Stewardship Initiative and the Michigan Virtual University. Her work has focused on designing effective and meaningful teacher professional development in the areas of science and mathematics.

DESCRIBING THE PROBLEM-BASED LEARNING PROCESS

As a science teacher, you probably use a variety of approaches and strategies in the classroom. On any given day you may lecture, lead group discussions, teach an inquiry-based lab, assign projects, ask students to complete individual reading and writing assignments, and perform many other types of tasks. All of these strategies have a legitimate purpose, and we encourage teaching that employs a diverse range of activities.

Why Problem-Based Learning?

In this chapter, we will discuss why problem-based learning (PBL) is one of the many tools you should keep in your teaching toolbox, ready to be used at appropriate times during your teaching. We will also give you some background information about how PBL was developed, background on how it works in a range of disciplines, and a basic framework for a PBL lesson. In later chapters, we will provide further detail on the "nuts and bolts" of teaching a PBL lesson and how to develop and facilitate learning activities using this strategy. The advice we will offer and the science problems we will share in later chapters come from our own experiences in using PBL to teach concepts to teachers. Many of the lessons have also been used with students across a wide range of age groups.

The reason we have used these lessons is because of a driving philosophy that it is imperative to help students develop the ability to inquire, solve problems, and think critically and independently (Barell 2010). Many of the thinking skills directly taught in the PBL process are included in the goals of 21st-century skills (Barell 2010; Ravitz et al. 2012). PBL is well suited to achieving the goal of developing thinking skills because it presents learners with authentic stories that require application of scientific concepts to construct and evaluate possible actions. In the process of solving problems, students plan, gather, and synthesize information from multiple sources or from investigation findings, evaluate the credibility of their sources, and communicate their ideas as they justify their claims. Students are guided by a set of simple prompts that help them organize information and generate questions and hypotheses.

In our experience, learners quickly adopt this framework as a habit of mind, and they begin to apply this critical-thinking strategy to other problems and real-world situations. The framework becomes a habit because the process is easily internalized and uses simple language. Asking the question "What do we know?" is easy for most students to remember and use, and the rest of the framework is just as direct and intuitive. This process also resembles KWL (McAllister 1994), a formative assessment strategy used widely in elementary classrooms. In KWL, students are asked to verbalize and record a list of what they "Know," what they "Want" to know, and what they have "Learned." The feature added by the PBL framework that makes it so "scientific" is the inclusion of hypotheses, leading students to make predictions and justify them.

Teachers in the professional development program for which these problems were developed very quickly adopted the language and turned "PBL" into a verb. When they encountered new problems, they initiated the process with phrases like "Let's PBL this." K–12 students are just as quick to adopt the cognitive framework. This is one of the benefits of using PBL in your teaching.

Historical Background of PBL as a Process

PBL's origins are rooted in this same desire to help learners solve real-world problems. PBL was originally a strategy for developing content knowledge in the context of assessing and diagnosing patients (Barrows 1980). Medical students had been very successful in memorizing information, but when asked to use the information to diagnose a patient, they were unable to apply their knowledge. What was lacking in their understanding was how the ideas they had memorized were useful in diagnosing and treating patients in an authentic "problem" they would encounter as a doctor. The challenge for medical school faculty members was finding a way to teach students to think like doctors, not like students preparing for a test. PBL also presents opportunities in such a contextualized manner, so medical schools began using this strategy. PBL was shown to be effective in helping medical students learn anatomy, pathology, and medical procedures and helping them apply this knowledge to medical cases. Thus, PBL became widespread in medical schools.

The same issues seen in the field of medical education are important concerns for science students, too. Just as second-year medical students struggle to transfer what they learn into practice, science students struggle to understand how memorizing metabolic pathways is helpful in explaining real-world issues, or how life cycles in different species are similar. Bransford and Schwartz (1999) suggest that transfer of knowledge is enhanced if the concepts are shown in a variety of contexts, rather than always presenting them in the same or very similar contexts. They also recommend using metacognition to support the transfer of knowledge across contexts. One of the strengths of PBL is that the framework we will present is a metacognitive structure—students are expected to be aware of what they know and what they need to know to solve the problem.

Bringing PBL to Other Disciplines

Since its beginnings in medical education, PBL has been adapted to business, law, law enforcement, and other subjects (Hung, Jonassen, and Liu 2008) and has been modified for science teaching (Allen et al. 2003; Gordon et al. 2001). Research by Hmelo-Silver (2004) suggested that PBL leads to increased intrinsic motivation of learners to become more self-directed. Another study reported that teachers who use PBL in their classroom teach more 21st-century skills (Ravitz et al. 2012).

In this book, the model presented for using PBL to teach science content has features similar to the PBL activities from other subjects, but it has been refined through research-based evaluation of the process when used for teaching science content in the PBL Project for Teachers (McConnell et al. 2008), as described in the next section.

The PBL Project for Teachers

The context in which the materials presented in this book were created was the PBL Project for Teachers, a National Science Foundation–funded teacher professional development program (McConnell et al. 2008).[1] The PBL Project was designed to accomplish several goals, including deepening K–12 teachers' scientific understanding, developing inquiry-based science lesson plans, and facilitating a form of reflective practice that applied the same PBL principles to the study of teaching.

In this program, K–12 teachers spent three days of a two-week institute learning science content surrounding standards they had identified as areas of need in their curriculum. Facilitators for each of the eight content strands planned PBL lessons to address those specific standards. These facilitators were experts in their respective science content areas who worked in teams of at least three. The teams wrote PBL problems that addressed the science standards teachers identified, and then shared these problems with peers for review. The problems were then tested and revised in an iterative fashion over four cohorts of teachers. The final versions were the basis for the problems found in this book and those to be included in future volumes in the PBL series.

The activities were modified for use in the K–12 classroom, with a focus on problems for life science, Earth-space science, and physics. These modifications included changing the context of the story to relate more to students in specific grade bands and changing the reading level to match the target audience. The concepts addressed in the problems remained consistent, in part because pre-assessments with teachers revealed very similar prior understandings as K–12 students, especially for teachers who were not science majors. Research to assess content learning showed that most of the teachers gained a deeper understanding of their chosen content as a result of the PBL lessons (McConnell, Parker, and Eberhardt 2013). Participants then used the content knowledge they gained to develop inquiry-based lesson plans and used PBL to analyze problems in teaching practice. Many of the problems have also been tested in K–12 and college courses, with revisions made to address any difficulties encountered.

---

[1] National Science Foundation special project number ESI-03533406, as part of the Teacher Professional Continuum program.

The PBL Framework

Throughout the PBL Project for Teachers, we used the same framework for designing and facilitating the PBL lessons that we used for the problems presented in this book. This framework draws from the guidelines described by Torp and Sage (2002) and the model used by the Michigan State University College of Human Medicine (Christopher Reznich, personal communication, October 11, 2004). In this model, students are presented with a problem, usually in the form of a story divided into two parts (Christopher Reznich, personal communication, October 11, 2004). There can be more than two parts to the story, but the key feature is that information is presented to students in stages. Figure 1.1 shows a representation of the PBL process.

Figure 1.1. The PBL Process

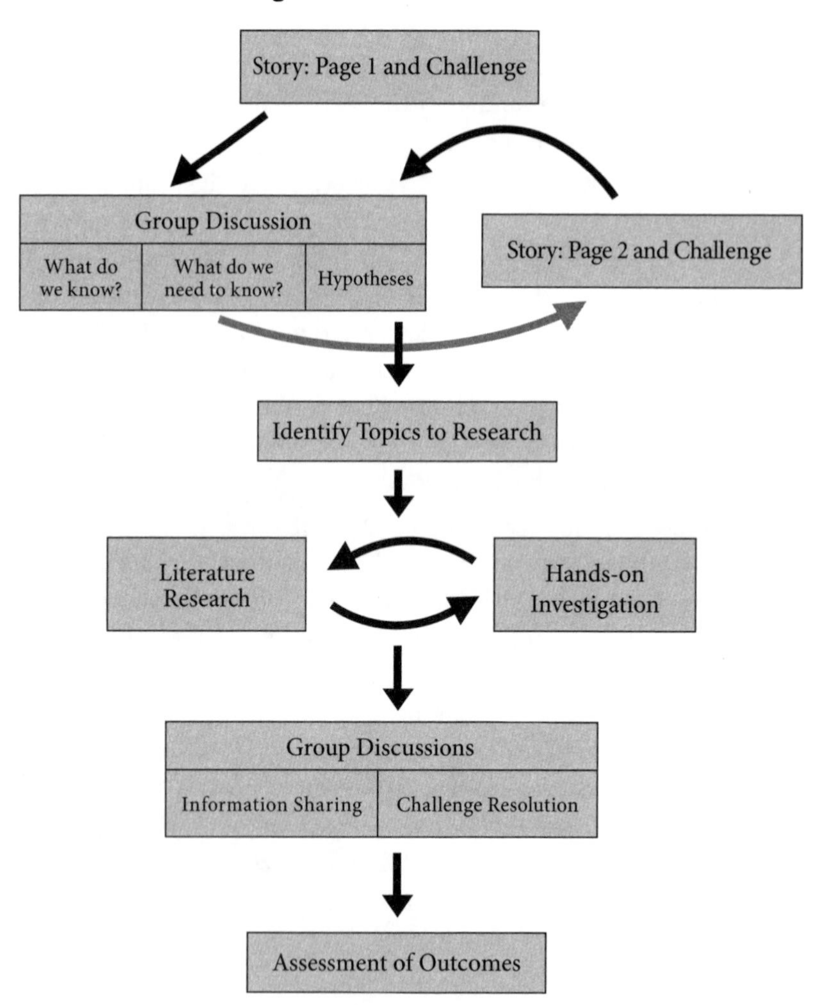

One of the keys to PBL is to develop a problem that is open-ended and ill-structured (Gallagher 1997; Torp and Sage 2002). By *ill-structured,* we mean a problem in which not all the information needed to solve the problem is presented to the learner, and some of the information presented may not be needed. In the real world, problems do not present themselves with a set number of variables or with a value for every variable provided but one. So the PBL lesson starts by presenting one of these scenarios and expecting students to unpack the problem and construct a path to solve the problem, as Dan Meyer (2010) describes for teaching math when he cites Albert Einstein: "The formulation of a problem is often more essential than its solution, which may be merely a matter of mathematical or experimental skill." PBL offers a way to engage students in thinking about the problem, developing a strategy to solve the problem, and exploring the content knowledge needed to achieve a solution.

Torp and Sage (2002) also use language to describe the scenario presented by the teacher that is common to most literature on PBL. They refer to the story as the *dilemma*. In the PBL Project for Teachers, we used the same language, but for this book, we have changed the terms. The origins of the word *dilemma* suggest that there are "two answers" to the scenario, but in reality a good dilemma is likely to have (at least at first glance) far more possible answers. So we have elected to call our scenarios *problems.* We will present the problems in the form of a story that ends in a *challenge*, which helps define the task students will take on and launches the research and analysis that follows. The challenge then serves as a focal point that defines what successful completion will require.

Structure of the Problem

In Chapters 5-8, we present PBL problems, which are the lessons that we have developed to share with teachers. Each chapter focuses on a content strand that fits within the life science discipline. These chapters begin with a description of the learning goals (Big Ideas) and the conceptual barriers students often face as they learn the concepts the problems address.

We have also included some interdisciplinary connections for each chapter. For elementary and middle school teachers, it may be helpful to demonstrate to administrators that students are developing more than just science content during the lesson. Our lists are by no means exhaustive, so feel free to make other interdisciplinary connections as well.

The pages that follow the opening discussion include the problems. We have developed a consistent structure for each problem to help you find and use the resources you need. The main components of each problem are listed below and briefly described in the following subsections; see Chapter 3 for more details on facilitating each component.

- Alignment With the *NGSS*
- Page 1: The Story
- Page 2: More Information

- Page 3: Resources, Investigations, or Both (not included in every problem)
- Teacher Guide
- Assessment

Alignment With the *NGSS*

The first page of each problem presents an overview of the problem and the alignment with the *Next Generation Science Standards* (*NGSS*; NGSS Lead States 2013). The table presented on this page lists the performance expectations, science and engineering practices, disciplinary core ideas, and crosscutting concepts that are addressed in the lesson.

Below the table is a list of keywords and concepts and a short description of the context of the problem. The items on this page are intended to help the teacher select problems that best suit his or her curricular needs.

Page 1: The Story

In the next component, a very brief story presents an authentic scene that sets the context for the problem, followed by a "challenge" statement making it clear what the learners are expected to accomplish. In our model, Page 1 is the "engagement" activity to draw students into the learning situation and show them a real-world reason to know some science concepts. The Page 1 story ends with "Your Challenge," a question or series of questions that presents the students with a problem they need to resolve.

This page is the starting point to a whole-class discussion of the problem using a highly structured analytical framework. This framework prompts students to generate ideas and questions within three categories: "What do we know?" "What do we need to know?" and "Hypotheses." Table 1.1 gives an example of the framework as it might appear when used in The Purple Menace problem in Chapter 6, "Ecology Problems." In Chapter 3, we present in more detail how the teacher facilitates this discussion, but one product of the initial discussion will be an analysis of the Page 1 story.

Table 1.1. Sample PBL Analytical Framework: The Purple Menace Problem

WHAT DO WE KNOW?	WHAT DO WE NEED TO KNOW?	HYPOTHESES
• Purple loosestrife has increased from 1 to 50 plants in 10 years. • Cattails live in the marsh. • The purple loosestrife is near the boat ramp.	• What is purple loosestrife? • What defines "successful" in plant populations? • Why is purple loosestrife a problem? • What eats purple loosestrife? Cattail?	• Cattails will be harmed if purple loosestrife spreads. • Purple loosestrife will attract new animals to the marsh, making the marsh more diverse.

Page 2: More Information

During the discussion of the Page 1 story, students will eventually exhaust their ideas and questions using the analytical framework described in the preceding subsection. When the generation of ideas slows down, the facilitator presents Page 2. This part of the story is sometimes longer than Page 1, and it gives students more details about the story that are needed to solve the problem. There may also be new unnecessary information, but students will again decide what is important during a second round of the analytical discussion. During this discussion, students' ideas are added to the same three columns generated from discussion of Page 1.The challenge is repeated at the end of Page 2 to help the students keep that focus.

Page 3: Resources, Investigations, or Both

During the second round of the analytical discussion, groups also need to decide what their priorities will be for doing further research. The class needs to decide which items in the "What do we need to know?" column to explore further, then create a plan for finding information to address those items. The teacher can ask students to search for information individually or in groups, set time limits, plan for access to a computer lab or the technology in the classroom, and define a plan for sharing information when the searches are completed. In some cases, this research phase may also include some hands-on investigations or field observations.

As groups move on to some small-group research, they may need help finding relevant information. The Resources page includes references and links that may streamline the search process so that groups can make the best use of class time. The purpose of this list is to help the teacher steer students toward these resources during the research phase of the lesson. In some problems, we list resources on Page 2 (More Information) as well as on a separate Resources page. When we do, this is intended to push students to use those resources as they analyze the problem and scaffold their analysis process. To find the resources easily, you can view the Resources Archive on the book's Extras page at *www.nsta.org/pbl-lifescience.*

Some problems include suggested investigations instead of or in addition to resources on Page 3. Each investigation provides a list of materials and a step-by-step procedure.

Once students have explored the resources or completed an investigation, the class convenes as a large group. In this final group discussion, students review the three columns in the analytical framework and present and discuss proposed solutions to the problem. There may be more than one appropriate solution to the problem, so students can justify and defend their ideas, using the evidence and resources they have collected to support their ideas.

Teacher Guide

The Teacher Guide includes the problem context, a model response, and, for some problems, an activity guide. This component of the PBL lesson provides resources for the teacher to use. The *problem context* gives you some background information about the scenario and science concepts associated with the problem, in case you are not familiar with the scenario. Each of the problems is aligned with standards, but this context goes a bit deeper and can help you explain the problem to students if needed.

The *model response* helps you assess the quality of responses your students generate. Just as you write down your ideal answer on an essay question before you grade it, we felt it would help to see what our content experts expect to see as a complete and accurate response. But keep in mind there may be more than one solution to the problem! The concepts included in our model responses could sometimes be applied in different and creative ways. Your students may not be able to provide explanations at this level, but knowing the explanation will help you assess how deeply your students understand the problem and its solution.

Some of the problems presented in Chapters 5–8 include hands-on activities that support students' learning. For these lessons, we have included an *activity guide* with instructions on implementing the activity. These activity guides will resemble lesson plans you might find in other resources, with materials, instructions for finding or constructing materials, and a description of the activity and the concepts it addresses. For example, in the cellular metabolism problem Why We Are Not What We Eat, the activity guide suggests having students test potatoes to see if they are respiring using an acid-base indicator in a glass or plastic container. The activity guide may also include links to other resources or diagrams you may need to copy as a template for the materials you will make for the classroom. In some problems, the activity guide will describe ways to incorporate the investigations listed on Page 3 of the problem into your lesson plans. Each investigation will have a Safety Precautions section, if appropriate, to address safety standards and practices.

Assessment

We have provided assessment items in a variety of formats to help you gauge student learning. The assessment items include transfer tasks, solution summaries, two types of open-ended questions for pre/post assessment (general questions and application questions), and common beliefs inventories. For each problem in Chapters 5–8, model responses are provided for the transfer task(s), solution summary, general and application questions, and Common Beliefs Inventory.

One of our goals is to help students apply concepts to new contexts. Transfer tasks can be used to assess that. Each transfer task is an open-response item in which the same concept addressed in the problem can be applied. For example, in the Baby Hamster problem in Chapter 5 students learn about the life cycle of a hamster, and the transfer task asks them to compare the life cycle of hamsters with the life cycle of other animals, including humans.

With each content strand, we have also included two types of open-ended assessment questions: general questions, which assess the breadth of knowledge about concepts; and application questions, which assess a student's ability to explain how the concepts apply to specific phenomena. In addition, there is a "Common Beliefs Inventory" assessment section with statements that reflect both accurate and inaccurate understandings; students are asked to answer "true" or "false" and to explain their reasoning with their answers.

Assessment is discussed in more detail in Chapter 4 ("Using Problems in K–12 Classrooms"), including the role of each type of assessment, the design of the assessments, and how they support teaching and learning in the PBL framework.

References

Allen, D. E., B. Duch, S. Groh, G. B. Watson, and H. B. White. 2003. Professional development of university professors: Case study from the University of Delaware. Paper presented at the international conference Docencia Universitaria en Tiempos de Cambio [University Teaching in Times of Change] at Pontificia Universidad Católica del Perú, Lima.

Barell, J. 2010. Problem-based learning: The foundation for 21st century skills. In *21st century skills: Rethinking how students learn*, ed. J. Bellanca and R. S. Brandt III, 175–199. Bloomington, IN: Solution Tree Press.

Barrows, H. S. 1980. *Problem-based learning: An approach to medical education*. New York: Springer Publishing.

Bransford, J. D., and D. L. Schwartz. 1999. Rethinking transfer: A simple proposal with multiple implications. *Review of Research in Education* 24 (1): 61–100.

Gallagher, S. A. 1997. Problem-based learning. *Journal for the Education of the Gifted* 20 (4): 332–362.

Gordon, P. R., A. M. Rogers, M. Comfort, N. Gavula, and B. P. McGee. 2001. A taste of problem-based learning increases achievement of urban minority middle-school students. *Educational Horizons* 79 (4): 171–175.

Hmelo-Silver, C. E. 2004. Problem-based learning: What and how do students learn? *Educational Psychology Review* 16 (3): 235–266.

Hung, W., D. H. Jonassen, and R. Liu. 2008. Problem-based learning. In *Handbook of research on educational communications and technology*, 3rd ed., ed. J. M. Spector, M. D. Merrill, J. van Merriënboer, and M. P. Driscoll, 485–506. New York: Routledge.

McAllister, P. J. 1994. Using KWL for informal assessment. *Reading Teacher* 47 (6): 510–511.

McConnell, T. J., J. Eberhardt, J. M. Parker, J. C. Stanaway, M. A. Lundeberg, M. J. Koehler, M. Urban-Lurain, and PBL Project staff. 2008. The PBL Project for Teachers: Using problem-based learning to guide K–12 science teachers' professional development. *MSTA Journal* 53 (1): 16–21.

McConnell, T. J., J. M. Parker, and J. Eberhardt. 2013. Problem-based learning as an effective strategy for science teacher professional development. *The Clearing House: A Journal of Educational Strategies, Issues and Ideas* 86 (6): 216–223.

Meyer, D. 2010. TED: Math class needs a makeover [video]. *www.ted.com/talks/dan_meyer_math_curriculum_makeover?language=en.*

NGSS Lead States. 2013. *Next Generation Science Standards: For states, by states.* Washington, DC: National Academies Press. *www.nextgenscience.org/next-generation-science-standards.*

Ravitz, J., N. Hixson, M. English, and J. Mergendoller. 2012. Using project based learning to teach 21st century skills: Findings from a statewide initiative. Paper presented at Annual Meetings of the American Educational Research Association, Vancouver, BC, Canada.

Torp, L., and S. Sage. 2002. *Problems as possibilities: Problem-based learning for K–16 Education.* 2nd ed. Alexandria, VA: Association for Supervision and Curriculum Development.

ALIGNMENT WITH STANDARDS

One of the issues all teachers need to consider when designing curriculum and planning lessons is the standards for their content area. The content experts who wrote the problem-based learning (PBL) problems presented in this book were very cognizant of the importance of standards and used standards to guide the selection and creation of the topics, the tasks included in the lessons, and the assessments included with each problem.

In fact, the list of problems developed for each content strand was created to address concepts aligned with state and national standards (NRC 1996). Each participant in the four cohorts of teachers in the PBL Project for Teachers was asked to list the top three choices of standards they felt needed the most development in their own curriculum. From these lists, the planners developed content strands to address concepts that would meet each teacher's needs.

The standards used in this process included the Michigan Grade Level Content Expectations (MDOE 2007) and the National Science Education Standards (NRC 1996), both of which were the relevant standards at the time of the PBL Project. Since then, a new set of standards has been published. The standards alignment presented in this chapter has been revised to address the *Next Generation Science Standards* (*NGSS*; NGSS Lead States 2013). As you use the problems in this book, you should also consider your state standards and local curriculum maps to guide your choices.

A Framework for K–12 Science Education and the *NGSS*

In 2012, the National Research Council released a document describing the structure of the new standards for science education, *A Framework for K–12 Science Education: Practices Crosscutting Concepts, and Core Ideas* (*Framework;* NRC 2012). The *Framework* laid out this structure as the foundation for the new standards that guide teachers to address more than just content. The PBL problems in Chapters 5–8 align well with the *Framework* because learners apply a variety of process skills and practices as they engage in specific science concepts, all of which fall within overarching themes that tie all the sciences together.

The *Framework* labels these different types of skills and concepts as the three dimensions of science learning. Each of these dimensions is directly connected to helping students relate to and understand any scientific phenomenon. The term *dimensions* is meant to connote that the concepts and practices should be learned and used simultaneously rather than consecutively in the pursuit of understanding. Teachers should use phenomena (rather than generalities) as the context in which students practice and develop the three dimensions of scientific literacy. As shown in Box 2.1, the three dimensions of the *Framework* are scientific and engineering practices (SEPs), crosscutting concepts (CCs), and disciplinary core ideas (DCIs). In the next three subsections, we discuss the ways in which these dimensions of the *Framework* align with and are expressed in the PBL problems included in this book.

Box 2.1. The Three Dimensions of *A Framework for K–12 Science Education*

- Science and engineering practices
- Crosscutting concepts
- Disciplinary core ideas

Each of the problems in Chapters 5–8 begins with an overview section that includes a table outlining the problem's alignment with the *NGSS*. Each table includes the performance expectations, science and engineering practices (this is the term used in Appendix F of the *NGSS;* NGSS Lead States 2013), DCIs, and CCs associated with the problem. To help you find problems that fit your needs, we have included a catalog of problems (pp. xiii–xiv) that lets you see at a glance which problems will help you teach specific concepts within their specific content strands. The catalog includes keywords and concepts and the grade bands that include related standards.

Science and Engineering Practices

One of the dimensions in the *NGSS* is SEPs. These practices describe skills and processes that scientists use in *doing* science, but they are more than just skills, so the authors of the *Framework* (NRC 2012) used the term *practices*. Most teachers will recognize these practices because the language incorporates the process skills and elements of the nature of science (Lederman 1999) expressed in the standards published by state departments of education. The SEPs included in Appendix F of the *NGSS* are showin in Box 2.2.

One of the benefits of PBL is that students are developing many of these practices as they progress through the analytical framework we introduced in Chapter 1 and explain in more detail in Chapter 3, "Facilitating Problem-Based Learning." If you present each problem in the format we have used, students take part in practices 1 and 4 during the discussion of the story; practices 3, 4, and 8 as they complete research on the problem; and practices 6, 7, and 8 during the final discussion of the solutions they propose. In some

Box 2.2. Science and Engineering Practices in the *NGSS*

1. Asking Questions and Defining Problems
2. Developing and Using Models
3. Planning and Carrying Out Investigations
4. Analyzing and Interpreting Data
5. Using Mathematics and Computational Thinking
6. Constructing Explanations and Designing Solutions
7. Engaging in Argument From Evidence
8. Obtaining, Evaluating, and Communicating Information

problems, the research component may include practice 5 as they find and process data, practice 3 if they conduct hands-on experiments, and practice 2 if they use models to explore or explain certain phenomena.

While we acknowledge that the practices are woven throughout each of the problems, we have also identified specific key practices that are strongly emphasized in each content chapter. These key practices are listed in the tables describing the alignment with the *NGSS*.

Disciplinary Core Ideas and Performance Expectations

Another dimension of learning in the *NGSS* is DCIs (the third dimension in the *Framework*). This is a very important dimension for teachers to consider in their lesson planning because the DCIs correspond to the content standards teachers must address. The *NGSS* present these ideas as statements of scientific ideas and labels them with a code that designates the content area, a DCI number, and the grade level.

As an example, one of the DCIs mentioned in Chapter 6, "Ecology Problems," is LS2.C-MS. This is part of the second (2.C) life science concept (LS) in the list of middle school standards (MS). The text of the DCI states the concept that students are expected to understand: "Ecosystems are dynamic in nature; their characteristics can vary over time. Disruptions to any physical or biological component of an ecosystem can lead to shifts in all its populations." In this statement, you should see a CC (Systems and System Models) and the science concept (ecosystems are dynamic and any change can affect other populations).

The *NGSS* also include performance expectations that are associated with the DCIs. The performance expectations are descriptions of indicators that students understand the DCI. These indicators are very helpful to teachers because they define the tasks and performances that can become both the activities and the assessments of learning we plan for the classroom. Using the same example from above, the performance expectation cited in Chapter 6 is "MS-LS2-2: Construct an explanation that predicts patterns of interactions among organisms across multiple ecosystems." The performance expectation describes a behavior that reflects one of the practices (construct an explanation) relating directly to the DCI about interactions among organisms. This same structure is reflected in all of the DCIs and associated performance expectations in the *NGSS*.

The performance expectation quoted above, along with the others cited in Chapter 6, were selected by the authors of the problems presented in Chapter 6 because teachers in the PBL Project for Teachers identified the concepts of changes in ecosystems and population ecology (MDOE 2007) that they felt they could strengthen in their teaching. The state standards correspond well with the *NGSS*, even though the newer standards were written 5 years after the state standards. We expect that you will find your state standards have a great deal of overlap with the *NGSS* as well.

Crosscutting Concepts

The *NGSS* identify seven crosscutting concepts—ideas that span across multiple science and engineering disciplines and are useful as a way for students to connect their understandings in an integrated fashion. These CCs are more than important concepts; they can be guides for making sense of new material. For example, in life science one of several important CCs is the flow of energy and matter. At a microscopic level, cells break down organic molecules to release energy for cell functions and must obtain or build those molecules. At the macroscopic level, energy and matter flow through food webs and biogeochemical cycles. Thus, the CC of flow of energy and matter helps students connect these seemingly unrelated phenomena. The CCs described in Appendix G of the *NGSS* (NGSS Lead States 2013) are shown in Box 2.3.

Box 2.3. Crosscutting Concepts in the *NGSS*

1. Patterns
2. Cause and Effect: Mechanism and Explanation
3. Scale, Proportion, and Quantity
4. Systems and System Models
5. Energy and Matter: Flows, Cycles, and Conservation
6. Structure and Function
7. Stability and Change

To help you align your teaching with the *NGSS*, we have explicitly identified the CCs expressed in the problems we have developed. Chapters 5–8 each contains a set of related problems addressing science concepts from a specific content strand (e.g., cellular metabolism in Chapter 8). As you consider your curriculum planning, look for the list of CCs at the beginning of each problem, in the table presenting the alignment with the *NGSS*.

Accurate Understanding in a Self-Directed Process

One of the concerns we have heard expressed by teachers relates to students' exposure to inaccurate concepts during the initial discussions of the problem. The PBL framework asks students to make a "What do we know?" list, and sometimes learners will make statements based on prior knowledge that might not match current understandings of science. For instance, participants involved in the genetics problems made inaccurate claims about the model of inheritance of sex-linked traits. Some teachers fear that hearing these misconceptions might be contagious or could plant seeds of ideas that we want to avoid.

But in reality, learners already come to the science classroom with their own ideas and conceptual understandings, regardless of whether we discuss them in class. The PBL framework uses the "What do we know?" list as a way to organize ideas, and it can be used as an assessment tool to reveal prior ideas. Rather than avoiding discussion of prior ideas, PBL offers a process to examine those ideas and find out if they align with evidence and scientific theories. The process allows learners to challenge and modify their understanding rather than merely introducing new and competing ideas to the individual's schema for understanding the world.

In the process of facilitation, our enactment provides a way for the teacher to handle concepts that are not scientifically accurate. Facilitators are encouraged to not reject learners' comments during the analysis that accompanies Page 1 and Page 2 of the story. This is a challenge when students present ideas that the teacher recognizes as inaccurate. To deal with this situation, we implemented a practice of establishing some guidelines for the analytical discussion. One of these rules is that any idea listed under "What do we know?" needs to be verified or confirmed by information from the PBL story or from another reliable source. Ideas only supported by "I've always heard that …" are placed under the "What do we need to know?" or "Hypotheses" column. Later, as students search for sources or carry out investigations, the group can revisit the lists and cross out those they find are not supported by theories, laws, and evidence. Through this mode of learning, students learn *why* some ideas are inaccurate rather than merely being told their ideas are wrong.

Chapter 3 discusses this issue in more detail, including this manner for guiding the discussion, but teachers using PBL should not shy away from discussion of inaccurate assumptions. The process of working through "wrong ideas" is essential in bringing about conceptual change. Accepting that their ideas may be disproved is one of the attitudes we want students to develop. In this way, PBL helps teach and support many aspects of the nature of science (Lederman 1999).

References

Lederman, N. G. 1999. Teachers' understanding of the nature of science and classroom practice: Factors that facilitate or impede the relationship. *Journal of Research in Science Teaching* 36 (8): 916–929.

Michigan Department of Education (MDOE). 2007. *Grade level content expectations*. Lansing: MDOE. *www.michigan.gov/documents/mde/SSGLCE_218368_7.pdf* (for K–8). *www.michigan.gov/documents/mde/Essential_Science_204486_7.pdf* (for high school).

National Research Council (NRC). 1996. *National Science Education Standards*. Washington, DC: National Academies Press.

National Research Council (NRC). 2012. *A framework for K–12 science education: Practices, crosscutting concepts, and core ideas*. Washington, DC: National Academies Press. *www.nap.edu/catalog.php?record_id=13165#*.

NGSS Lead States. 2013. *Next Generation Science Standards: For states, by states*. Washington, DC: National Academies Press. *www.nextgenscience.org/next-generation-science-standards*.

FACILITATING PROBLEM-BASED LEARNING

The experience of being the teacher in a science classroom during a problem-based learning (PBL) activity is a bit different than what you might experience for other types of lessons. In some learning activities, your role is that of content expert or presenter of information. The students might be involved in recording information, listening, or perhaps applying new ideas. Alternatively, students might be carrying out some kind of science investigation as you direct and guide with questions. These roles are certainly appropriate, but PBL requires something different.

In PBL, the teacher definitely steps away from the lead role and instead becomes a *facilitator.* Educators use this term a lot in teaching, but for our model of PBL, we believe this role is accentuated. The facilitator's role is to provide minimal information but to provide resources and ask questions to guide the process. The students become more active participants in the discussion and even take the lead in identifying next steps and issues that need to be explored.

These new roles take practice—for both teacher and student. Students need to take risks in sharing ideas and "defending" their ideas using information and evidence. Your role requires skillful questioning to guide without leading, and just as important, the ability to say nothing and let students explore their own ideas to find their misconceptions. In this chapter, we will use a vignette format to provide examples of what you might see in a classroom in which PBL is being taught, with a focus on how the teacher can guide discussions during the lesson. We will also share tips and strategies for successful facilitation of a PBL lesson; additional tips are provided in Chapter 4, "Using Problems in K–12 Classrooms." Some of what we share in this chapter is the result of our research on effective facilitation of PBL (Zhang et al. 2010), and some is based on our personal experience and teaching styles.

Remember, as you implement the lessons you select from this book, you may find that you need to practice your role as a PBL facilitator, and it takes time and practice to learn how to respond to students' ideas on the fly.

Moves to Make as You Go Along: Stage-Specific Advice

Facilitating PBL problems feels very different from traditional teaching and may require some strategies that are not part of your normal routine. Throughout this chapter, we will offer some "moves" you can plan to make. These are deliberate tactics to help your students think and talk about the problem they are analyzing, and they help you move into facilitator mode. It can be hard to remember that your role has shifted. You need to hold in some of your expertise and let your students struggle a bit with the challenges of solving a real problem. It is hard to do this, because you want to help them, but in the long run, stepping into the role of facilitator will help your students gain confidence and skills they need to think critically. And that's an important goal!

At the same time, there are times when the teacher needs to share his or her knowledge of the concept. This may mean giving some examples of phenomena that demonstrate a process, or explaining how certain ideas are connected. The teacher also may need to ask questions to informally assess students' understanding or clarify what a student means by a comment or question. These moves are important in facilitating students' analysis of a PBL problem and in helping students make sense of the information they are finding. Part of the art of facilitation is learning when to use your content knowledge and when to hold back and let students explore an idea. For the beginning facilitator, we recommend patience: if in doubt, let students work for a bit, and then share your expertise.

Explaining Discussion Guidelines

Because you and your students may be experiencing PBL for the first time, it is important to set some guidelines for a PBL lesson. Discussion about real-world problems may reveal some strong opinions, some misconceptions, and some differences in beliefs and values that may be difficult for younger learners to understand. Before you start a PBL lesson, at least until your students learn to operate in this new type of lesson, setting some guidelines will help you manage the discussion and keep the conversation on task and respectful.

In the first section of the vignette, Ms. Sampson shows the class a list of guidelines for discussing PBL problems. These guidelines are useful in creating a climate in which participants are able to share ideas, pose questions, and propose hypotheses. They may also help create a culture of open discussion in your classroom. Throughout the vignette in this chapter, we have tried to indicate when the science and engineering practices (SEPs) and the crosscutting concepts (CCs) from the *Next Generation Science Standards* (*NGSS*; NGSS Lead States 2013) appear in this lesson. See Chapter 2, "Alignment With Standards," for a complete list of the SEPs and CCs.

Ms. Sampson's Science Classroom: Discussion Guidelines

Ms. Sampson has been planning since the summer to try a new lesson idea. Today, she's starting a PBL activity that she thinks will take about three days for her seventh-grade science class to complete. The topic is invasive plants in her "Ecology" unit, and today's activity follows some readings about ecosystems and food webs and a video about marsh ecosystems.

Ms. Sampson: Class, today we're going to start learning about a type of plant we can find in this area. As we work, you will take the role of a committee that manages the natural area at Rose Lake. We are going to use PBL to look at this topic, so we need to set some discussion guidelines.

She projects a slide with the guidelines and discusses the list (see Box 3.1).

Box 3.1. Guidelines for Discussion

1. Open thinking is required—everyone contributes!
2. If you disagree, speak up! Silence is agreement.
3. Everyone speaks to the group—no side conversations.
4. There are no wrong ideas in a brainstorm—respect all ideas.
5. A scribe will record the group's thinking.
6. The facilitator/teacher will ask questions to clarify and keep the process going.
7. Support claims with evidence or a verifiable source.

Helping Students Function in a Self-Directed Classroom

This recap of discussion guidelines is important to help students start to manage their own learning. Although the PBL framework introduced in Chapter 1 is a good foundation for critical thinking, students may not have experience using a structured process for solving problems. In essence, we are making the metacognition needed to support learning more explicit (Bandura 1986; Dinsmore, Alexander, and Loughlin 2008) in a process that will help students develop the type of self-directed learning abilities we hope all our students can achieve.

The guidelines are important in helping students develop the habits of scientific discourse. A conversation in a scientific context is different from a conversation with friends about sports, music, politics, books, or other topics. So to help our students learn to function in a scientific community, or even just to understand the process behind scientific claims they might read about in an online news source, they need to know how we share and develop ideas in science.

At the same time, the guidelines are a reminder to the facilitator about his or her role in the discussion. As the facilitator, one of the most difficult tasks is avoiding the urge to give "right answers" to your students. But it is important for you to set an example by respecting new ideas or ideas you are uncertain about. Your role, especially at the beginning of a PBL problem, is to ask questions to clarify, to solicit responses from students who may be hesitant to share ideas, and to be the "referee" when the class rejects one student's ideas before any evidence has been discussed.

Recording Information

In the guidelines that Ms. Sampson shares, she mentions the "scribe." It is important to have a durable record of the ideas students generate. The written copy of the ideas students generate is also important as a "map" that students and the teacher can follow to see the development of their understanding. In a sense, posting the ideas as a list makes the learning "visible." The facilitator will use this list to make choices about guiding questions, information search strategies, and activities that can support the type of learning each particular class needs.

In some cases, you may wish to have a student serve as the scribe, but this may pull that student out of the conversation. It is difficult to create or share your own ideas when you're busy writing others' ideas on the board, and your students are probably not able to juggle those tasks. In our experience, it is best if you, the facilitator, can

> **Technology Tip**
>
> SMART boards (interactive whiteboards) and similar technology are a good option for recording group discussions! They allow you to record a "page" of notes, move to a new page, and return to previous notes when needed.

record students' statements, questions, and hypotheses on the board or projected on the screen so all students can see the lists (see Figure 3.1).

Figure 3.1. Recording Learners' Ideas in the PBL Framework

You can create areas on the whiteboard for each of the three categories of ideas in the PBL framework ("What do we know?" "What do we need to know?" "Hypotheses"), but we suggest you use large pieces of paper taped to the board or the wall. This will let you add pages as the students' list of ideas grows. You can make notations or cross off statements and hypotheses as the students find new information, but it is important to have those items to look back at during the process of working through the problem. Students can see how their understanding develops, question why they think an idea is true, and connect the evidence with their new understandings. The large pieces of paper or electronic files will also allow you to move back and forth between different sections, if you teach the subject more than once per day.

Launching the Problem

Once you have established discussion guidelines and procedures, it is time to launch the problem. For this stage, you can have students arranged in small groups, seated on the floor in a circle, seated in desks, or whatever arrangement works best for you.

In Chapters 5–8, each PBL problem begins with an overview that describes the key concepts of the problem and aligns the problem with the three dimensions of the *NGSS* (NGSS Lead States 2013). This alignment includes a table describing the SEPs, disciplinary core ideas, and CCs addressed in the lesson. Keywords and a context for the problem are also offered to help you identify the problems that are most appropriate for your curriculum.

Following the overview and alignment page, each problem includes the text for "the story" arranged in two parts. Page 1 is the part of the story you will use to launch the activity. Most of the stories are short and can be printed on a half-sheet of paper. In some cases, you might project the story on the screen, but we find that it is helpful to give each student or group a hard copy so they can refer to it as they work through the analytical framework. You may choose to print one copy per student or let pairs or small groups read from the same page.

Start by handing out the copies of Page 1, and ask your students to read the story quietly. You might need to make accommodations for special needs students. Once everyone has had time to read through the story, ask one person to read the story aloud. This may seem redundant, but it is actually a very important step. Our research has shown that groups that read both silently and aloud at the start of the story generate a significantly higher number of ideas, questions, and hypotheses than groups that only read the story silently. We posit that in the first reading, students are working to comprehend the story, and in the second one, they begin forming their own ideas in their minds. The time to process the story and think quietly seems to be important in supporting the discussion in the group as they move forward. The vignette sections that follow provide examples of what this process looks like in the classroom setting.

Ms. Sampson's Science Classroom: The Launch

Ms. Sampson: OK, class, today's PBL is called The Purple Menace. Here is Page 1. Please read this story quietly. I'll give you about 2 minutes.

She hands out Page 1 of The Purple Menace problem. (See Chapter 6, p. 140, to read the story.) As her class reads, she tapes three large pieces of paper to the board, labels them "What do we know?" "What do we need to know?" and "Hypotheses," and gets her colored markers ready. After 2 minutes, she asks for a volunteer to read the story. David volunteers, stands, and reads the story aloud.

Ms. Sampson: Thanks for volunteering, David. Now that you've heard the story, let's look at our three categories on the board. What do we know about the story right now?

The class is quiet for a minute, but she notices that the students look like they are thinking.

Andrea: Purple loosestrife has increased from 1 to 50 plants in 10 years.

Ms. Sampson writes Andrea's comment on the "What do we know?" paper.

Jamal: There are cattails in the marsh too.

Marcus: And it says they are found by the boat access. I think they mean the boat ramp.

Mai: What is purple loosestrife? I've never heard of it.

Ms. Sampson: Mai, I can add that to the "What do we need to know?" page. Good question!

David: This says the loosestrife is successful, but I'm not sure what that means.

Carmela: I think they mean that they are increasing in number. They've gone from 1 to 50.

Ms. Sampson: OK, we can put that under "What do we know?" but the question goes under "What do we need to know?"

Carmela: Yeah, I think we need to look that up.

Moves to Make: "Unpacking Ideas"

During a discussion in the three-column framework described, students are very likely to bring up terms and concepts that need to be "unpacked." *Unpacking* is a term commonly used in education and business conversations, but it is not always clear what unpacking an idea entails. In essence, students are using one of the SEPs as they analyze and interpret the information they are given (see SEP 4, p. 13). Students also use this stage to define the problem (see SEP 1, p. 13).

Let's focus on an example from the preceding vignette section. Marcus brings up an idea to include in the "What do we know?" column:

Marcus: And it says they are found by the boat access. I think they mean the boat ramp.

Mai: What is purple loosestrife? I've never heard of it.

Ms. Sampson: Mai, I can add that to the "What do we need to know?" page. Good question!

Until the discussion begins, the teacher may be uncertain what students know about the scenario described in the story. Knowing what this plant looks like and where it lives is certainly important to the problem! But it is clear from Mai's question that not all the students in the class are familiar with it or know why it is relevant to their local ecosystems. Ms. Sampson steers this comment to the "What do we need to know?" list and moves on.

It may be easy to imagine a discussion of purple loosestrife later in the lesson, but another useful strategy would be to "unpack" the concept of *purple loosestrife* right away. This can be done with questions that draw on what the students know about it already. These questions could be asked during the initial discussion, or they could wait until the class starts to explore the "What do we need to know?" list in more detail. But there are a couple of different ways to handle unpacking the concept.

Let's compare the teacher-as-expert approach with the teacher-as-facilitator approach (see Table 3.1, p. 24). In the "expert" role, the teacher shares what she knows, and the students become passive recipients. In the "facilitator" example, Ms. Sampson pulls information from the students, and the students' role shifts to either the expert or the problem solver who recognizes the need to find information. In the latter example, the students are active learners and consumers of ideas, a role we want students to master.

In the facilitator example, the students get much of the same information, but they have either discovered or remembered the information on their own and in their own words. The students have begun to develop some independence in learning and are practicing the skills used by proficient problem solvers. Independent learners can do more than just recall and repeat ideas. They synthesize ideas from information they are given or collect themselves (SEP 4: Analyzing and Interpreting Data). To demonstrate deep understanding, students should be able to synthesize information by connecting ideas in the context

Table 3.1. Teacher-as-Expert Approach Versus Teacher-as-Facilitator Approach

TEACHER AS EXPERT	TEACHER AS FACILITATOR
Mai: What is purple loosestrife? I've never heard of it. **Ms. Sampson:** Mai, I can add that to the "What do we need to know?" page. Good question! Purple loosestrife is a plant found in wetlands across the Midwest. It grows about 6 feet tall, and spreads quickly when it gets in the marsh. **Denise:** So why does that matter? **Ms. Sampson:** Well this plant is not originally from this area. And some people are worried about what it will do to the plants normally found here. **Mai:** Oh! So they might kill other species off. I think I get it. **Ms. Sampson:** OK! Here's a picture of it so you'll know what we're talking about.	**Mai:** What is purple loosestrife? I've never heard of it. **Ms. Sampson:** Mai, I can add that to the "What do we need to know?" page. Good question! OK, does anyone know what purple loosestrife is? **Devin:** I'm not sure, but I'm guessing it's a plant. They're talking about cattails, and those plants are really common here. **Denise:** Is it OK if I do a search for *purple loosestrife* on the web?" **Ms. Sampson:** Sure! We'll come back to your answer in a few minutes. **Marcus:** My dad takes us fishing there all the time. I think I've seen those plants. They're kind of tall, and they have purple flowers at the top. Is that the right plant? **Ms. Sampson:** I think they're the ones we're talking about. **Denise:** Here it is! There are some pictures here. It says purple loosestrife is a Eurasian plant that lives in wetlands and was introduced to the United States in the 1800s. **Mai:** OK, that helps. Maybe we can move the purple loosestrife into the "What do we know?" section.

of a real problem, instead of repeating bits of disconnected facts. In the expert example, Ms. Sampson is hinting toward the concept of competition, but we only see evidence in the facilitator example that Mai and Marcus have started making a connection between the concept and the problem.

Generating Hypotheses

As students work through the analysis discussion of Page 1, they are likely to state ideas that reach beyond "What do we know?" and "What do we need to know?" In the next section of the vignette, watch for the comment that suggests an inference. Sometimes these are subtle, but as the facilitator, you can point out the step the student has made and suggest adding this new idea to the list of "Hypotheses."

As a facilitator, you will need to pay attention to the questions students ask during the discussion. One common pattern is that learners will present an idea as a question when they have some uncertainty about the statement. A student may suggest a question to add to the "What do we need to know?" list, but the question is actually a tentatively worded hypothesis. Let's look at an example of this.

Ms. Sampson's Science Classroom: Generating Hypotheses

Ms. Sampson: OK, class, we've figured out what loosestrife is. Do you have any other things we need to learn about or ideas about this problem we should add?

Angie: I have a "need to know" thing. Does it matter that there are cattails there? I'm guessing the cattails are important, because the thing Denise found online said that loosestrife spreads really fast. I bet the cattails are going to be hurt by this.

Ms. Sampson: Good question, Angie. You're thinking about the marsh as a system now [CC 4: Systems and System Models]. But I think I hear a hypothesis in that statement. You're asking if there's an interaction between loosestrife and cattail, but if we reword that, can we make this a hypothesis?

Angie: I'm not sure if I'm right, though. I'm not sure this is a good hypothesis.

Ms. Sampson: But that's OK, Angie! Remember, a hypothesis is a proposed answer to a question that can be tested, and if the evidence eventually shows that it's not correct, that's alright! So do you want to try to build a hypothesis from your question?

Angie: I guess so. I'm not sure how to start it, though. "I predict cattails and loosestrife will … " Is that the way to state it?

Carlos: Shouldn't we use the same kind of words we use in other labs? *If, then,* and *because?*"

Ms. Sampson: That's what we use when we're going to change a variable and see what the result is, Carlos, but that's a start. Who remembers what we use when we're observing events instead of changing a variable?

Allysa: Isn't that when we use the "I think that … " kind of hypothesis?

Ms. Sampson: Yes, that's right, Allysa! So, Angie, use that as a start. "I think that … "

Angie: OK. "I think that loosestrife will take over the marsh and cattails will die off."

Joseph: It needs a "because" statement.

Ms. Sampson: Yeah, what would be the *because* part?

Angie: "Because plants introduced from other places usually outgrow native plants."

Ms. Sampson: Good! That's our first hypothesis.

Jason: Wait! Can we really say they do that? We don't know for sure!

Mrs. Sampson: That's right, Jason, but this is a hypothesis. We'll try to check to see if this really happens, and if not, we can reject a hypothesis.

David: No, that's not what happens. I don't think that's right.

Ms. Sampson: Remember, we're making hypotheses. We need evidence before we can reject a hypothesis, so I think we need to include it on the "Hypotheses" page.

Carlos: I have a different hypothesis. I think different animals eat loosestrife, so the animals will change, but the area around the boat ramp will still be pretty much the same.

Ms. Sampson: You need to put it in hypothesis form, too!

Carlos: How about this? "I think the loosestrife will attract new animals, but the marsh will be unchanged otherwise, because having more kinds of plants just means you can have more kinds of animals."

In this example, a student initiated the first hypothesis, but it began as a "What do we need to know?" question. Note the way that Ms. Sampson directed the discussion toward the "Hypotheses" column in the analysis discussion and pointed out that Angie's question seemed to include a hypothesis. This is a very common pattern in the discussion of Page 1 with most problems, and you need to watch and listen for those types of questions. One cue is to look for a "because" statement in the question. For instance, if a student says, "I want to know if the cattails will die off because the loosestrife takes over," this suggests a hypothesis. The "because" indicates a connection between cause and effect (CC 2: Cause and Effect: Mechanism and Explanation) or a rationale for a possible solution to the problem (SEP 6: Constructing Explanations and Designing Solutions). The teacher could easily leave the question worded as it is, but it helps to move it to the "Hypotheses" column. Students can then "test" the hypothesis as they do information searches later in the lesson.

The strategy Ms. Sampson used was to point out the purpose of a hypothesis and mention that the question asked sounded like a testable question. She then asked students to rephrase the question rather than doing the rephrasing herself. This puts more control of the process in the hands of the students so they must practice this skill. Ms. Sampson is truly taking the role of facilitator by steering students with questions and letting the students generate the final version of the hypothesis. This facilitating includes reassuring Angie that it was fine to hypothesize and later find that the hypothesis is not supported. You've probably seen students' reluctance to be "wrong" on a hypothesis, and PBL helps them get over that fear.

It helped that Ms. Sampson's class had learned a deliberate pattern for writing hypotheses in other classes. If you have been working on SEP 3 (Planning and Carrying Out Investigations), your students will likely have begun learning this skill as well. In your class, part of the scaffolding you will do with students is to help them learn to ask questions, write hypotheses, build data tables, and write explanations. PBL gives you yet another context in which students can use those same practices, so you have the flexibility to insert your particular format for structuring these elements of the science process.

Angie's hypothesis took quite a bit of scaffolding. Students contributed bits and pieces and made connections with the class "standard" for hypothesis writing. It was not an automatic process at first. This is typical of students who are still learning to think like scientists. Carlos was able to phrase his hypothesis in the appropriate format much more quickly because he was part of the process of working out that format during the discussion about Angie's hypothesis. This is also a common event. Students very quickly adopt the structure when the class works through the process out loud and can see the hypothesis on the list as a reference for later discussion.

If no students come up with hypotheses on their own, the teacher needs to help students think about making some predictions or proposed solutions. As the list of "What do we know?" and "What do we need to know?" items grows, a facilitator can ask something like, "So what do you think is the answer to the challenge at this point?" This is usually enough to get the ball rolling with the first hypothesis.

Our experience suggests that once the first hypothesis emerges, other students become more comfortable suggesting possible solutions or hypotheses. In other cases, students may need a prompt from the facilitator. You can elicit hypotheses by asking, "So what do you think is the answer to the challenge?" or "Do have any hypotheses about a solution?" If students are really having trouble framing an initial hypothesis, you can ask if they think there is a relationship between any of the things listed under "What do we know?" Defining relationships is often the beginning of a hypothesis. Such initial hypotheses may not be complete answers to the challenge, but they start the ball rolling.

Introducing Page 2

As your students work through the PBL analysis framework and the information on Page 1, there will be a moment when the students start to run out of new ideas to put in the three categories of the framework. They will exhaust the "What do we know?" ideas and address most of the learning issues on the "What do we need to know?" page. The list of hypotheses might be short, but the generation of these ideas will slow down. *When that happens, your job as the facilitator is to transition to Page 2.*

Page 2 continues the Page 1 story and adds new information that will help students work toward a solution to the challenge statement at the end of Page 1. Introducing Page 2 should work very much the way introducing Page 1 did; students will read Page 2 quietly, then a student will read it aloud. Once that happens, the class can repeat the analysis process adding new ideas to the same three categories of the PBL framework.

One major difference in the way to handle information relates to the new content in Page 2. You may find that "What do we need to know?" items on your chart will be answered with the Page 2 story, or that the hypotheses generated in the first discussion will be rejected based on the new information. You can certainly add new questions and hypotheses as well as "What do we know?" statements, but we strongly recommend that you keep the first set of ideas on the board and visible to students. As you answer items in the "need to know" list, cross them out but leave them on the chart. Some facilitators keep a list of "summarized knowledge" under each question to connect the "need to know" items with the new information they use to answer the questions. When you learn enough to eliminate a hypothesis, don't delete or erase it, but cross it out. Having those ideas visible is helpful when students look at the path they have taken from their initial ideas to the final solution for the problem. Processing their own ideas this way gives students a way to know *why* the solution works, not just that this is the right answer. It also builds a habit for students to show their thinking and their work. You might even find that when students begin to adopt the PBL skills as habits, they apply them in other subjects as well!

Ms. Sampson's Science Classroom: Introducing Page 2

Jason: OK, so we've mentioned pretty much everything in the story, but I still don't understand why having loosestrife in the area is a problem.

Ms. Sampson: So do you want to put that under "What do we need to know?"

Jason: Yeah, I think so.

Ms. Sampson: Okay, got it. What else can we add to our lists? (long pause) Any other ideas? Or new hypotheses? (long pause)

Ms. Sampson: OK, then it sounds like you're ready for more information, right?

(Multiple students): Yeah! We need more information.

Ms. Sampson: Alright then, here's Page 2. Let's do what we did with Page 1. Read the story to yourself, and then we'll read it out loud.

She hands out Page 2, the class reads it quietly, and Devin reads Page 2 aloud.

Ms. Sampson: OK, good. Now let's add new pages for "What do we know?" "What do we need to know?" and "Hypotheses." We need to talk about each of these pages again with the new information we have. So ... what do we know NOW?

Will: Cattail is a keystone species.

Rose: Yeah, but what does THAT mean?

Marcus: Yeah, I want to know too! We need to put *keystone species* under "need to know."

David: It says cattails support a diverse ... what is it? "A very diverse wetland community."

Marcus: Yeah, but is that connected to this keystone thing?

Ms. Sampson: OK, it's under "need to know."

Tricia: We know the lake was made by a glacier.

Jason: There's something about rhizomes. I don't know that word.

Ms. Sampson: Alright, another "need to know!"

Denise: That last part of the story says again that the loosestrife is from Eurasia, and it says it has caused problems in other areas.

Andrea: But it doesn't say what kind of problems. I think we need to know more about that.

Ms. Sampson: Good point! Let's put that under "need to know," too.

Denise: This thing about rhizomes ... it says cattails are going to eliminate the ecosystem. Does that mean the lake will go away?

Vince: Cattails can't make a lake go away.

David: But they spread by the wind. Have you ever seen those fluffy things blowing away?

Jordan: Yeah, and milkweed does that, too! [CC 6: Structure and Function]

Ms. Sampson struggles to let the conversation work its course—they are getting off track and starting to talk about issues that are not important to the problem.

Allysa: Hey! Maybe loosestrife has those fluffy seeds, too. It spreads fast and the wind could do that. I bet cattails spread the same way. [CC 1: Patterns]

Mai: Never mind that! The story asks about how to reduce the spread of loosestrife. I think we need to look that up. I think they should spray the loosestrife to get rid of it.

Ms. Sampson: OK, Mai, let's talk about that last part of Page 2. We need to think about how to control loosestrife. If you think it's important, do you want to make that a hypothesis? Vince, we can add your idea as a hypothesis too.

Mai: Yeah! I think spraying loosestrife with weed killer will keep it from spreading because it will die off.

Sarah: I don't think that will work. Wouldn't that kill the cattails, too? I think you have to dig them up and get rid of them, because using chemicals might hurt other things.

Ms. Sampson: Let's put both of those under "Hypotheses."

Carmela: Yeah, those are good ideas, but we still need to know more about cattails and loosestrife. And if there have been problems in other places, we might be able to look that up.

Ms. Sampson: We could, and you'll get a chance to do that soon.

Moves to Make: What If Students "Go Down the Wrong Path"?

In this section of the vignette, we see Ms. Sampson guiding the class through the analysis phase of Page 2. Students listed the new ideas they got from Page 2, raising questions about ideas they didn't understand and offering new hypotheses. But we also see an example of students "going down the wrong path." Some conversations take off on tangents, like the comments about the cattail seeds, and others may follow incorrect hypotheses that the teacher knows are going to lead to a dead end.

As the teacher, you will encounter those moments when you want to comment to prevent the class from following a "wrong" hypothesis. You should already know what some viable solutions to the problem are, and you simply want to help your students find the right answers. But it is important *not* to interject comments that stop students' exploration of incorrect ideas. A hypothesis that is later rejected is a powerful learning experience and is likely to lead to enduring understandings. So you need to let students explore those ideas, even when your instincts tell you to steer them in a new direction. Teachers are likely to want to correct the inaccurate ideas right away, but the PBL framework emphasizes letting students find evidence that leads them to eliminate ideas on their own.

Note how Ms. Sampson handled it. She allowed the class to work through their ideas, and she included Vince's hypothesis in the list. You should avoid eliminating hypotheses for your class. Let students decide when an idea is rejected. That's a difficult thing for teachers to do, and it may take some practice, but it is important! In this case, Sarah helped the process by introducing a new hypothesis to compete with Mai's. Including them both will allow students to compare them, using evidence and information they collect. Eventually, the students will have all the tools they need to decide which is the most viable hypothesis.

When you encounter this type of situation, be assured that it's normal in the PBL process. Each of the authors has experienced this, and we have felt the same internal conflict between providing content knowledge or letting students learn or discover for themselves. We've all learned to be patient, let the students drive the discussion, and wait for the learners to see all the information before we simply give answers.

But there are good strategies for redirecting the discussion! One suggestion is to establish a practice in which you, the teacher, are free to participate as a learner. This gives you permission to ask the same type of questions students should be asking. In this co-learner role, you can model critical thinking and questioning while using your comments to keep students on task and on track.

Here are some questions or statements, or "steering tools," you can use to keep your class discussion on track:

- "So how does that apply to the challenge for this problem?"
- "Maybe we should restate the question we are trying to answer."
- "Do we have a source that can verify that idea?"
- "What kind of evidence do we need to support that?"
- "How does this info from Page 2 relate to Page 1?"
- "That sounds like a 'need to know' issue."

Researching and Investigating

Once your students have completed the discussion of Page 1 and Page 2, you should have an extensive list of items under the three categories in the PBL framework: "What do we know?" "What do we need to know?" and "Hypotheses." On some of the pages, you may have crossed out questions you've answered or hypotheses you've ruled out. The information that is left should point to learning issues and predictions that have potential as solutions to the challenge presented in Page 1. Remember, the goal is to propose solutions to the challenge, so the research and investigation should focus on this goal.

The next step in the process of facilitation is to help the class develop a plan for gathering information or conducting an investigation that will answer the "What do we need to know?" questions that are still unresolved. In this phase of the PBL process, the teacher has some choices that will determine what the next part of the lesson will include. Is there an inquiry-based lab or hands-on investigation that would help students understand the concepts that underlie the problem? Will students use a computer lab or classroom computers to search for information on the internet? Are there text resources that can help them answer the questions? Should the teacher provide a limited set of readings to ensure that students find productive information? All of these may be appropriate choices!

Investigations

In some problems, there may be a hands-on activity, such as a model that students can build, that would help illustrate a concept. For instance, in Chapter 6, "Ecology Problems," the Bottom Dwellers problem is an ideal situation in which to use water sampling from a local stream or lake. By measuring dissolved oxygen using a simple test kit, students can learn more about how we test bodies of water to determine what might be causing a decline in some species of animals. This allows students to experience a real-world phenomenon and use data as one type of evidence they can use in constructing their final solutions.

You may also have inquiry-based investigations your students can conduct to learn or reinforce specific concepts. Cellular metabolism problems such as Mysterious Mass may include a lab investigation to burn a twig. An elementary life cycles problem like Wogs and Wasps may be supported by observing mealworms or tadpoles in the classroom over a period of a few weeks.

One of your roles as the teacher is to plan for these investigations. You may have activities in your textbook resources that would be appropriate, or you may find or create new lab activities to meet your needs. In Chapters 5–8, we have provided some lab activities that fit with specific concepts, including instructions to help you plan and implement these activities.

An important component of any activity is safety. Students and teachers need to learn how to properly assess risks and take actions to minimize risks. Safety issues to consider include the use of sharp objects, the safe use and disposal of chemicals, and the presence of fire hazards. Teachers are responsible for taking precautions such as wearing safety goggles or glasses, providing disposal containers for sharps and chemicals, and ensuring that students know where fire extinguishers and chemical showers are located.

Information Searches

Other problems are best addressed by helping students search relevant resources for answers to the learning issues they have identified. For teachers who need to integrate literacy standards into science teaching, the skills of finding and evaluating information from multiple sources are clearly featured in this part of the PBL process.

Sources for answering the learning issues your students have identified may include web searches, their science texts, books in the school's library, or magazines and newspapers. Although our first thoughts seem to turn toward technology as the go-to source, many text-based tools are certainly appropriate. You can decide which tools are best suited for the context in which you are teaching based on access, convenience, or the "fit" for the topic at hand.

The search for information also offers multiple choices for scheduling. Perhaps you will have students work on this the same day they have analyzed Page 1 and Page 2, or you may need to plan this phase for the next day or as homework. The number of days you spend on this task also depends on your specific needs.

Ms. Sampson's Vignette: Beginning the Information Search

Ms. Sampson: OK, class, you've created a good list of facts, hypotheses, and things we "need to know." Now we need to plan what information we'll look for next. Let's look at the "What do we need to know?" list. Are there specific ideas that groups will offer to find out more about?

The students talk softly with their groups about what they want to research.

Jamal: Our group wants to look up cattails and what they mean by keystone species. Can we look for that?

Ms. Sampson puts Jamal's name next to cattails.

Ms. Sampson: You got it, Jamal! Your group can get started.

Denise: We'll look for stuff about loosestrife, like what eats it.

Ms. Sampson: OK, Denise, that's a good topic to look at.

Jason: What about the problems loosestrife caused in other places? Can we look that up?

Ms. Sampson: Sure! Do you three want to look that up?

Jason: Yeah, we'll take that topic.

Rose: There's a hypothesis about how to keep it from spreading. Someone needs to look that up.

Ms. Sampson: Good idea! If you'll volunteer, you can do that. David, how about if your group helps Rose's by looking up the weed killer topic. It's related to Rose's group.

David: OK, we can do that.

Ms. Sampson: Alright then, folks! You need to get started with the time we have left today, and we'll continue working on this tomorrow.

Angie: Can we look stuff up at home tonight, too?

Ms. Sampson: Sure! But make sure you keep a record of what sources you find, and bring it with you tomorrow. Remember, when we're done, each group is responsible for sharing what you find with the entire class. Be organized!

Teacher-Selected Sources

For some classes, "searching" for information may require more assistance from the teacher. In these cases, the teacher might pick a limited collection of resources and provide these resources to groups when they are ready to find answers to their learning issues. Perhaps the problem is complex enough that you want to steer students to specific sources. Maybe the information they need is not easily accessible to your students, either because very little is published online about the topic or because your school filters access to the necessary sites. Even the age or technology skills of your students may suggest that you should preselect the sources.

One strategy for doing this is to create sets of articles or websites that address specific topics. You can either give each group of students all of the sets or distribute each set to a different group. The latter option forces students to read and analyze the texts and share what they find with other groups. This type of communication is common among practicing scientists and addresses skills that students need to develop across the curriculum.

To help you select problems for which preselected sets of sources are useful, we strongly recommend that you work through each problem in advance. Think of the types of "need to know" issues you expect students to identify and try searching for those concepts. If you can't find them easily, your students may also struggle to locate sources. Many of the problems in Chapters 5–8 include a Resources page (Page 3) with links to websites and references to other materials that are relevant to the science concepts.

Sharing and Resolving the Problem

When your students have completed the investigation or information search, the next phase includes sharing what they found. If each group has selected specific learning issues to research, this sharing is critical to the challenge presented to the class. No one group is likely to find all the information they need to solve the problem or build a complete solution to the challenge. But if they share information, the class can co-construct some solutions, as project teams do in the workplace. This phase of the PBL process gives students a chance to hone their skills with SEPs 6–8: Constructing Explanations and Designing Solutions; Engaging in Argument From Evidence; and Obtaining, Evaluating, and Communicating Information.

The class sharing session should still focus on the three pages of analysis they created during the discussion of Page 1 and Page 2, especially the "What do we need to know?" and "Hypotheses" pages. The information search should address specific "need to know" items, and their findings should help in the evaluation of some of the hypotheses as they apply what they have learned to the challenge presented in the story. Post the three pages on the board or on a wall for all to see and take a minute to recap what the class has done so far.

Each group should be asked to share. While some students may be reluctant to speak in front of the class, building their comfort with such a task is an important learning goal. We find that when the presentation is informal, the task is less threatening. One way to promote sharing is to ask a student in each group to share one thing they learned. This leaves room for others in the group to share their ideas. Sharing their findings also helps students learn to pay attention to evidence and reliable sources.

As groups present what they found, it may also help to have other students take notes or record concepts in a journal or science notebook. They should also be encouraged to ask questions that help clarify ideas. Let your class know that the goal is not to stump or quiz each other, but to help the entire class understand the information.

If your class or specific groups did an investigation, this is a good time to have the class look at the procedures and results and talk about what the evidence means. If you have a standard procedure for presenting scientific explanations from an investigation, this is a perfect time to apply that structure. For instance, you can establish a procedure in which students share observations and data, identify patterns in the data, and suggest an explanation for the patterns. In the case of developing a solution for a problem, another approach is to describe the proposed solution, explain why it will work, and explain how evidence supports the ideas. If you have a structure you use for this in your current lab activities, you can use the same structure with your PBL lessons.

When all the information has been presented, you have options on how to construct solutions. One way to come to a final answer to the problem or challenge is to discuss the problem as a group. The focus on this should be the hypotheses created by the class. When a group wants to support a specific hypothesis, you can ask for a rationale: What evidence makes you think this is a good hypothesis? Other students should also be allowed to make counterclaims about a hypothesis or to present ideas that would refute the hypothesis or solution. This discussion can be a rich assessment of students' learning and ideas because it forces students to reveal the connections they make between concepts as they apply them to an authentic problem. Recording their ideas may be helpful if you wish to assess these connections, or you may choose to have a checklist so you can keep track of evidence of new learning.

In some of the classrooms in which we have observed teachers using PBL lessons, we have also seen another approach. Some teachers elect to have each group talk about the evidence they have found and create their own solution to the problem. This works best if each group was responsible for looking up more than one concept from the "What do we need to know?" list. It is helpful to set a time limit for this discussion, and you may want to have a structure for the group's response as described earlier in this section. The teacher may also have a handout with general questions for the group to answer. This can include what hypothesis the group was investigating, what "need to know" issue they explored, what evidence they collected through research or experimentation, and how the evidence leads to a solution. The group then presents their ideas to the class, and other groups are encouraged to ask questions or explain what they see as problems in the solution.

In both of these scenarios, the next step is to ask for a solution to the challenge listed at the end of Page 1. This is the ultimate goal of the activity, so make sure you pay attention to the challenge. Students might present more than one solution. That's okay! In the real world, there may be multiple ways to solve a problem, and we want students to understand that. But when more than one solution is presented, you can ask the class to discuss the strengths and weaknesses of each solution, ask them to vote on the one they prefer, or ask each student to write a short response or exit ticket with a prompt similar to the following: "Which solution do you think is the most useful? Explain why you chose this solution over the others." (See the "Assessing Learning" section, pp. 40–41, and the "Responses to Assessment Data" section, p. 42, for more information on exit tickets.)

Ms. Sampson's Science Classroom: Sharing and Building Solutions

Ms. Sampson: Today, we're going to share the information you found about The Purple Menace problem we've been working on. As you present, remember that you need to describe the answers you found clearly, and you should be ready to tell us where you found them. We'll use that information to see what we can cross out on the "need to know" list and how your information fits with our "Hypotheses" list. I need each group to share what they found. Jamal, I'd like your group to start, if you don't mind.

Jamal: OK. We looked up cattails. We found a lot of pages with stuff about cattails. There's a PDF file that we liked from a good source that had a lot of information, but there were a bunch of other pages with a lot of the same stuff. It says they like shallow water and that they like to have what they called "wet feet." And they grow 3–10 feet tall. Some other pages talked about cattails being a big food source for a lot of animals. Birds, muskrat, insects, and some other animals eat them, but many other animals use them for nesting and shelter. They mention a lot of kinds of birds, muskrats, ducks and geese, rabbits, and even deer. The roots are in the water, so fish and insects hide in them, too.

None of these talked about keystone species so we looked that up, too. A keystone species is one that is really important for an ecosystem because so many other species depend on it. So we think that means the cattail is a keystone species because a lot of animals use it for food and shelter. We think the cattail is really important for the marsh or the edge of the lake.

Ms. Sampson: OK, that's a good start! It helps us understand how cattails are part of a system [CC 4: Systems and System Models]. We need to think about the role of the cattail as we learn about purple loosestrife.

Angie: Well then, I think our group should go next. We looked up loosestrife.

Angie, Denise, and Mai share information they found about loosestrife, including their origins in Eurasia, their ability to grow in shallow water, the way they spread their seeds by dropping them in water, and the fact that in North America, they are spreading fast because there are no animals that eat it. They found websites describing how marshes are being taken over by loosestrife, leading to the loss of species that depended on native marsh plants.

Ms. Sampson: Good information, girls! Let's think about what that means. How does loosestrife compare to cattails? We need to see if there are any patterns that might be important in this problem [CC 1: Patterns].

Jason: They live in the same kind of places—shallow water.

Carlos: Yeah, but nothing eats the loosestrife. Just about everything eats the cattail or lives in it.

Ms. Sampson: So what does that mean if they both end up in the same place?

Andrea: That must be why the loosestrife takes over the marshes.

Carmela: That's bad for the animals that need the cattail, isn't it?

Ms. Sampson: Good question! What do you all think?

David: I thought the loosestrife kills off the cattails.

Tricia: How can one plant kill another? I don't think that's what's going on here. Ms. Sampson, can you explain this?

Ms. Sampson: Well, let's look at what we already know! I don't think anyone is suggesting that loosestrife kills the cattail, but let's think about competition. Are the two kinds of plants using the same resources?

Jason: They live in the same part of the lake. They both like shallow water.

Rose: And they're both close to the same height.

Andrea: So what? They don't eat, so they're not fighting over food! How can they compete for food?

Ms. Sampson: Think about it. How do plants get food?

Marcus: Photosynthesis. They use sunlight.

Ms. Sampson: Right! Remember, we talked about the flow of energy through the ecosystem a couple of weeks ago [CC 5: Energy and Matter: Flows, Cycles, and Conservation]. Do you think they compete for sunlight?

Mai: Compete for sunlight? Why? There's plenty of it, right?

Ms. Sampson: But what if one shades out the other?

Angie: And our website said the roots of the loosestrife are so tightly woven that other plants can't grow there. I think the plants are competing for space.

Carlos: Oh, I think I get it! Since the loosestrife doesn't have anything eating it, it has an advantage, and it crowds out the cattails. But does that mean this would take a long time before the cattails are gone?

Ms. Sampson: Maybe! According to the story, how long ago did the loosestrife appear?

Carmela: It was from 1996 to 2006. That's 10 years, and there are only 50 plants. It's going to take a long time.

The class agrees that this helps them understand why loosestrife is a problem. Jason's group reports that they found a description of similar ideas on some websites from Minnesota and Wisconsin. Those states have passed laws banning the sale and distribution of purple loosestrife. They also share reports that a single loosestrife can produce thousands of seeds each year, and that once one plant grows in a marsh, the marsh might be all loosestrife within 8–10 years.

Ms. Sampson: Alright, we understand the problem. Now let's talk about controlling loosestrife.

Moves to Make: Correcting Misconceptions or Nonscientific Solutions

When your students are constructing and selecting solutions, they are considering information their class has shared, but they also are influenced by prior knowledge. Sometimes this prior knowledge is not accurate, and it is likely to be durable and difficult to change. These ideas can lead to solutions at the end of the analysis process that are not practical, fail to really solve the problem, create other problems, or omit concepts the teacher has identified as an important learning goal.

So what should you do when that happens? Our first suggestion is to assume the role of a classmate by asking questions you know will force the class to think about an important concept or piece of evidence. When skillfully used, these kinds of questions can help students notice the problems with their claims. One of the most effective approaches is to have students compare a problematic claim to information they have listed under the "What do we know?" column of the analysis charts.

One of the strategies that can be very effective is to ask questions like "Are there any of the 'what do we know' statements that contradict this solution?" In the vignette, Ms. Sampson asked students to review the facts they know about what resources each of the two plants needs as a way to get them to think about competition. By asking students to compare their researched information with other facts and evidence, you can help them develop SEP 8: Obtaining, Evaluating, and Communicating Information. This is a critical practice in our world of abundant information. Students will be exposed to many claims and proposals in the news, at work, through advertising, and in legislative bills that need critical analysis against the available evidence. This also helps address at least two of the "Essential Features of Classroom Inquiry" listed in the National Research Council supplement to the *National Science Education Standards* (NRC 1996, 2000), by asking students to give priority to evidence as they form and evaluate explanations.

Another approach would be to ask students to list the strengths and weaknesses of each solution. As in the strategy above, this places students in the role of evaluators and requires comparison of solutions to evidence. This also models the type of analysis used in the workplace for problems related to science and engineering, as well as many other contexts. Remember, the phase of the PBL process in which students generate solutions highlights both synthesis and critical thinking, so having students engage in these types of thinking is important.

But what if this doesn't do away with a misconception? Or what if the class didn't grasp a key concept that makes a big difference in the problem? Scientifically incorrect ideas can be durable and may get in the way of students' assimilation of new ideas. Some of the peripheral information may draw students' attention as they create solutions. So the teacher needs to be prepared to correct ideas and guide the development of solutions during this final part of the PBL lesson.

When your students just aren't applying concepts accurately, you now have a chance to explain ideas. There are times when your students need you to be the expert. Although we suggest you be patient with students' own thinking process, you may need to step in and present information that students need. If needed, you can lecture, lead a discussion, show a simulation or an image on the screen, or introduce some type of activity to help guide the learning. A good example of this is illustrated in the vignette when Ms. Sampson brought up the topic of competition. Competition is a key concept in understanding how invasive species influence an ecosystem, but none of her students had searched for information about competition. Direct teaching has its place in the classroom, and your content expertise is important. Ms. Sampson could use this opportunity to present a short lecture about competition for natural resources, or maybe use the time to do a short computer simulation as a supporting activity.

Assessing Learning

When implementing a PBL lesson, the teacher or facilitator needs to respond to the learning needs of his or her students as they emerge. Flexibility is key, but to be flexible the teacher needs information about what students are thinking. Assessment is an important part of the facilitation process. As you lead a class through PBL problems, you should be planning to assess and planning to use the information from your assessments to adjust your teaching.

The PBL process as we have described provides for continuous assessment. The process of analysis using the PBL framework allows the teacher to hear and see what students are thinking as they talk about their ideas and record information, questions, and hypotheses under the three columns of the analysis structure. Each comment from a student gives you insight into their understanding.

But be aware that what you hear in a group discussion may not reveal what every individual is thinking. In a whole-class discussion, the teacher sees a "group think" picture of what students know. There may be bits of information from a handful of students that seem to make sense when the entire group shares ideas, but you need to know what each student understands. It is helpful to have strategies that let you assess individual students rather than the entire group of students.

The need for individual assessments is even more pronounced if the activity takes more than one class period. As we developed our model in the PBL Project for Teachers, our facilitators found it very helpful to implement informal assessment strategies like exit tickets. These are very brief prompts asked before the end of a class period for which students write a short response. These prompts may focus on one idea the students learned, one idea they found confusing, or one question they have based on what happened in class. You might also ask students or groups to give a written summary of the information they found during their research, their choice of the "best hypothesis so far," or a drawing of the concept they are exploring.

Another form of assessment is the transfer task. "Transfer" of knowledge refers to the ability of students to apply knowledge of the concept in new contexts. For instance, knowing that tadpoles change shape as they develop into frogs is important, but students need to recognize that all living things have their own patterns of development. The importance of transferring knowledge to new situations is supported by Schwartz, Chase, and Bransford (2012), who suggested that a deep understanding of a concept must be accompanied by transfer. To help you perform this type of assessment, the problems in Chapters 5–8 include transfer tasks. The transfer tasks are often used as a summative assessment, but they can also inform the choices the teacher makes about the next activities to include in a unit.

In Chapters 5–8, we also present open-response questions that we have developed and tested for each content strand. There are two types of these questions—general and application—to address the concepts and standards included for the problems in the content strand. We discuss more about the role of these assessments in Chapter 4, as well as options for when to use the assessments and how to interpret responses.

Responding to Assessment Data

Assessment of learning is important, but you also need to consider how you can use the assessments to respond to students' needs. We've introduced a couple of assessment strategies that can help you select your next moves as a facilitator in the PBL lesson. But it may help to share some examples. These examples include exit tickets and group summaries of solutions to PBL problems.

Exit Tickets

This is a simple and quick way to collect information about your students' understanding and issues that need to be resolved. Exit tickets can ask one of several different kinds of questions, including "What's one thing you've learned?" "What about today's topic still confuses you?" or "What's one question you have about today's lesson?" (Cornelius 2013) Each student then writes a short response and turns it into the teacher at the end of class. The next step is for the teacher to read through the tickets to see if there are important issues that need to be handled in the next day's class. The vignette section that follows provides an example of how this might work in Ms. Sampson's class.

Ms. Sampson's Science Classroom: Exit Tickets

Ms. Sampson asked her class to write exit tickets after Page 2, using the prompt "one question you have about The Purple Menace problem."

Ms. Sampson: OK, I looked over the exit tickets you wrote yesterday, and I think we need to add something to the "need to know" list. Several of you wrote that you don't understand how loosestrife got into the area at Rose Lake. Can we add that to our list?

The class agrees, so this is added to a list of topics to be researched. Another possible result might be the following:

Ms. Sampson: Your exit tickets tell me that there may be some questions about the beetles that Denise's group found. Before we go on, let's talk about that idea and how the beetles help control the loosestrife.

She explains that biologists would test the beetles in a controlled lab space and then in a larger test plot before releasing them in the wild. She also reports that the beetles don't kill all of the loosestrife, but they control the population enough that cattails survive in the marsh, allowing the animals who depend on cattails to also survive.

Group Summaries

In the PBL Project for Teachers, we found that an entire class may agree to a solution, but some individuals may have a different level of understanding of the concept. One of the strategies we tested proved to be useful—group summaries.

In this assessment, each group is asked to write a summary of their group's proposed solution. The summary should include a description of the solution they think best solves the problem or answers the challenge along with a rationale that explains what evidence they used to construct their solution (SEPs 6, 7, and 8). In the process of discussing and writing this summary, group members are able to solidify their understandings. When groups are asked to complete a summary, individual scores on content tests are often higher than if the summaries are not used.

The following vignette section offers an example of how this assessment might be implemented in Ms. Sampson's lesson.

Ms. Sampson's Science Classroom: Group Summary of Solutions

Ms. Sampson asked each group to write and turn in a summary of the solution they had developed on the second day of the lesson. The groups were scheduled to share their ideas on the third day. In these summaries, Ms. Sampson noticed an issue that needed to be explained. Her students thought that purple loosestrife had the ability to "kill off" the cattails. Because this is a common misconception about invasive species, Ms. Sampson decided to address this in class.

Ms. Sampson: Alright, kids, I saw in your solutions that many of you used phrases like "kill" or "kill off" to describe what the loosestrife does to the cattails. I think we need to explain that a little bit more, so we're going to do a computer lab. The computers at your lab stations have a program called Pond Ecology, and we need to do the "Competition Activity" listed in the main menu. When everyone is done, we need to discuss your results.

The students worked in groups on the computer simulation that lets them introduce two species of fish that eat the same type of food and a predator that eats only one of the fish. They recorded data about population growth in their lab journals and then discussed the data in a large-group discussion.

Ms. Sampson: So what did you find out about the effect of competition in the pond?

Sarah: We noticed that both species of fish have bigger populations if the other is not around. Since they both eat the same food, they only have enough to keep a certain population alive. If someone else is eating the same thing, both have problems getting enough, and the populations go down.

Carlos: We saw the same thing. It makes a lot of sense. But the cool part was when we added a predator that ate just one of the fish, that fish's population went way down, and the population of the other fish went way up.

Ms. Sampson: Did the fish that went down disappear completely?

Carlos: No, not completely. There were still some fish, but not many.

Ms. Sampson: OK, that's good. Now let's think about how this relates to the loosestrife and cattail. What do they compete for?

Jason: Light. We talked about that earlier.

Mai: And space. Remember what we found out about the roots of the loosestrife crowding out other plants.

Angie: Ohhh … I think I get it! They both have lower populations because they compete, but because there's nothing that eats the loosestrife, it had a big advantage in the competition!

Ms. Sampson: That's a great connection! Now, let's go back to the computers and see what happens with both fish with the predator that eats just the first fish, and then add a predator that eats the second fish. That should help us see what the weevils and beetles do to help control purple loosestrife.

Summary

Facilitating PBL requires a slightly different set of skills than direct teaching, and it requires practice. Your role as the facilitator means you need to be prepared for several possible paths students may take. Your role also shifts from provider of information to a guide who needs to skillfully ask questions that allow students to reveal their own thinking, resolve their own misconceptions, and base their own ideas on evidence rather than an "expert" source. This questioning also requires you to moderate disagreements and keep students on task, so facilitating PBL lessons will feel very different than other lesson formats.

You will also need to anticipate what kinds of information, models, and explanations you should be ready to offer your classes. If you teach multiple sections of the same class, each may have very different needs, so you will find yourself selecting different responses. Assessment is a key factor; you need to know what your students are thinking! Box 3.2 presents some tips to remember as you facilitate your PBL lessons.

Box 3.2. Do's and Don'ts of PBL Facilitation

Do ...

- Use open-ended prompting questions.
- Count to 10 or 20 before making suggestions or asking questions.
- Allow learners to self-correct without intervening.
- Be patient and let learners make mistakes. Powerful learning occurs from mistake making. Remember that mistakes are okay.
- Help learners discover how to correct mistakes by clarifying wording, seeking evidence, or checking for discrepancies between ideas and evidence.

Don't ...

- Take the problem away from the learners by being too directive.
- Send messages that they are thinking the "wrong" way.
- Give learners information because you're afraid they won't find it.
- Intervene the moment you think learners are off track.
- Rush learners, especially in the beginning.
- Be afraid to say, "That sounds like a learning issue to me" instead of telling them the answer.
- Rephrase learners' ideas to make them more accurate.

Source: Adapted from Lambros 2002.

References

Bandura, A. 1986. *Social foundations of thought and action: A social cognitive theory*. Upper Saddle River, NJ: Prentice-Hall.

Cornelius, K. E. 2013. Formative assessment made easy: Templates for collecting daily data in inclusive classrooms. *Teaching Exceptional Children* 45 (5): 14–21.

Dinsmore, D. L., P. A. Alexander, and S. M. Loughlin. 2008. Focusing the conceptual lens on metacognition, self-regulation, and self-regulated learning. *Educational Psychology Review* 20 (4): 391–409.

Lambros, A. 2002. *Problem-based learning in K–8 classrooms: A teacher's guide to implementation.* Thousand Oaks, CA: Corwin Press.

National Research Council (NRC). 1996. *National Science Education Standards.* Washington, DC: National Academies Press.

National Research Council (NRC). 2000. *Inquiry and the National Science Education Standards.* Washington, DC: National Academies Press.

NGSS Lead States. 2013. *Next Generation Science Standards: For states, by states.* Washington, DC: National Academies Press. *www.nextgenscience.org/next-generation-science-standards.*

Schwartz, D. L., C. C. Chase, and J. D. Bransford. 2012. Resisting overzealous transfer: Coordinating previously successful routines with needs for new learning. *Educational Psychologist* 47 (3): 204–214.

Zhang, M., M. Lundeberg, T. J. McConnell, M. J. Koehler, and J. Eberhardt. 2010. Using questioning to facilitate discussion of science teaching problems in teacher professional development. *Interdisciplinary Journal of Problem-Based Learning* 4 (1): 57–82.

USING PROBLEMS IN K–12 CLASSROOMS

In Chapters 1–3, we described the design of problem-based learning (PBL) lessons, shared tips for facilitating PBL activities, and gave you a taste of what PBL looks like in the classroom setting. But you may still have questions about when to use PBL, how to integrate PBL into the rest of your curriculum, and how to manage students and groups as they work through the problems you present. In this chapter, we will share some tips based on our own experiences and on the experiences of other K–12 teachers in the PBL Project for Teachers. These tips will help you integrate PBL into your curriculum, identify potential resources, weave existing hands-on and inquiry activities into the PBL lesson, and assess student learning.

Establishing Continuity Across Lessons

Many classroom teachers like to establish a consistent set of routines or procedures for students to follow. This can help students learn habits of organization, acceptable ways to communicate, and skills needed to succeed in the classroom. If your classroom is built on specific procedures, PBL may feel very different than "the usual" routine. Students may take some time to learn the PBL process and how to think about problems in the PBL framework.

One of the benefits of PBL is that it helps students create the same kinds of productive habits of mind used by scientists in most fields. The three-category analytical framework used to discuss Page 1 and Page 2 creates a structure that students can apply to a wide range of different problems and content subjects.

But what if the PBL framework does not match your routines? Perhaps you like to start a unit with vocabulary and discuss concepts before doing an inquiry activity. Or maybe you like to start with a lab, discuss concepts, and then apply the idea to real-world problems. Many teachers have a set structure for phrasing hypotheses or an organization scheme for recording information in a lab notebook or science journal during labs.

For all of these examples, PBL can be incorporated as part of the routine. In fact, learners quickly adopt the analytical framework as a standard way to solve problems. PBL is also flexible enough to let you integrate other routines into the PBL framework. For instance, your students can express their hypothesis statements during the PBL analysis using an established format you use with other investigations. If journals are a part of your classroom routine, you can have students record their ideas in their journals as their groups discuss a problem. Or you may wish to have students post their "need to know" questions on a class wiki or cloud storage space to be answered as they find pertinent information.

We believe the key is to be consistent in some of the aspects of your teaching. As a professional, you can make choices about which procedures you maintain. You have the ability to modify parts of the PBL process to fit your teaching style and preferred routines.

It is possible you may opt to use PBL to make major changes in your teaching practice. Or you may be looking for new strategies to supplement your units. In the following sections, we describe some issues and strategies related to modifying PBL to fit in your curriculum.

Incorporating Other Activities

As you consider adopting problems in this book, you may be reluctant to give up the related learning activities you currently use. Maybe you have a great demonstration of cellular respiration that has students act out chemical reactions, or perhaps you have a lab activity in which your students measure the calories stored in peanuts and marshmallows. If these are effective lessons, you probably don't want to toss them out to implement a new activity from this book.

Some critics of PBL suggest that the problems presented in this format are not suited to hands-on investigations. Many of the problems described in this book are based on finding information from text and online resources, so this may be the case for some problems. But we have also included activities with investigations. We encourage you to insert your best inquiry activities wherever and whenever they fit! PBL *can* be inquiry-based teaching! As you read through this chapter, you will see some examples and possible strategies.

Let's look at the Bottom Dwellers problem from Chapter 6 of this book. The story describes Loon Lake in Michigan, where the population of mayflies has declined dramatically, and a neighboring lake that has not seen the same loss of the mayfly population. Page 2 of the story (p. 124) mentions that Loon Lake is surrounded by homes with septic tanks, while the homes around the neighboring lake have city sewer connections. This hint should lead students to research the effect of septic tanks on water quality—something they might be able to do as an investigation.

Investigation in the "Research" Phase of PBL

After reading and analyzing Page 1 and Page 2, students should have a list of "need to know" issues. They will plan a strategy for searching for information, which is likely to include searching the internet. Students can find the information they need to solve this problem entirely online. But this may also be an ideal time to plan for a lab activity. Information from other sources is great, but primary information collected by doing an experiment or investigation is a powerful thing. Because one of our goals in teaching science is to help students use evidence to support their claims, doing experiments should be part of your science teaching. Teachers can then help students relate the evidence they collect to the information found in online searches or from texts and help students construct solutions to the problem that reflect the conclusions drawn from these investigations.

A likely fit in the Bottom Dwellers problem would be water testing. Many schools have test kits or probes that can test dissolved oxygen (DO) or perhaps a stream nearby that has mayfly and midge larvae, which are described in the story. A natural modification to the lab is to have students collect or test water samples from a stream or lake to learn how to measure DO. Your class may be able to take samples of the macroinvertebrates in a stream or a wetland next to your school to see how scientists collect information to solve this real problem.

Modification Ideas for Bottom Dwellers Problem

The story relates to a decline in the mayfly population in a lake (see Chapter 6 for the complete story), and students are asked to investigate why the insects are disappearing. They can find answers by researching issues related to septic tanks at homes around the lake. Some modification ideas are as follows:

- Insert a lab to test dissolved oxygen in local lakes and to sample benthic macroinvertebrates.
- Insert an experiment to show the loss of dissolved oxygen from bacterial decomposition in samples with and without organic materials in the water.

Investigation as a "Teachable Moment"

Sometimes the lab activities you use make a great response to a "need to know" item that arises during a class discussion. Using the Bottom Dwellers lesson as an example, you might have students share information about decomposition consuming a lot of oxygen in a lake. A student may say, "I don't understand how tiny little bacteria can use that much oxygen! That doesn't make much sense to me." This might be the right time for an experiment! Perhaps you call these moments an opportunity for "just in time" instruction. PBL is certainly flexible enough to allow you to take advantage of these moments. For the Bottom Dwellers example, you could have students measure the biological oxygen demand by measuring DO in two samples of water, one with very little organic matter and the other with a lot of organic matter. Then let the samples sit in a dark cabinet for 48 hours and test the DO again. This is a very effective way to show students that bacteria are using up oxygen in the water when organic wastes are dumped in a lake or stream.

These activities might already be part of your curriculum. Rather than replace them with a PBL problem, a little planning will let you integrate your existing activities with the new PBL problems. In fact, giving students a chance to collect real data and use it with the PBL problem strengthens students' understanding of how science is important to solving authentic problems.

Making Time for PBL and Timing PBL

Teachers often ask about when PBL should be incorporated into an existing unit. Is it a way to introduce a topic, or should it be used to assess understanding at the end of a unit? Is it best used as an application task, or can students actually learn new concepts during a PBL lesson? The simple answer is … Yes to all of these! PBL is a strategy that is flexible. Figure 4.1 illustrates a number of ways in which PBL can fit in your unit plans.

Figure 4.1. Integrating PBL Into a Curriculum Unit

<table>
<tr><th colspan="7">DIFFERENT SEQUENCES FOR INTEGRATING PBL INTO A SCIENCE UNIT</th></tr>
<tr><td>PBL to engage learners</td><td>PBL activity: Engage students and introduce the relevance of a topic</td><td colspan="2">Discuss solutions, reinforce concepts</td><td>Investigation—inquiry activity with concept</td><td>Student report or presentation</td></tr>
<tr><td>PBL to explore a concept</td><td>Pre-assessment of a concept</td><td colspan="2">PBL activity: View the concept in context, build new ideas, confront misconceptions</td><td>Discussion for concept building</td><td>Summative assessment</td></tr>
<tr><td>PBL to apply a concept</td><td>Introduce a concept</td><td>Direct teaching—vocabulary and concepts</td><td>PBL activity: Apply the concept</td><td>Activities for practice, remediation</td><td>Test or project as summative assessment</td></tr>
<tr><td>PBL as a summative assessment</td><td>Introduce a concept</td><td>Lecture and reading from text</td><td>Practice—worksheets or sample questions</td><td>Lab activity</td><td>PBL activity: Use as application or summative assessment</td></tr>
</table>

Problems such as those presented in Chapters 5–8 can be used in many different ways in a unit. Often, teachers think of PBL as a follow-up to concept-building activities or as a summary activity to assess students' ability to apply the concepts taught during a unit. PBL certainly can be used in this way. As you read through the problems we have developed, you are likely to see clear contexts in which students can apply ideas you teach in other activities. When used at the end of a unit, students can demonstrate their ability to connect the new concepts to previous ideas or to integrate skills taught in other subjects. The groups' list of ideas generated during the analysis process ("What do we know?"

"What do we need to know?" "Hypotheses") can reveal the learners' ability to think critically, recognize the concept within a context, and use vocabulary appropriately to talk about scientific principles, data, and their inferences.

But the end of a unit is not the only time when PBL is a useful learning tool. Teachers can use the problems as an "engagement" or motivation activity to introduce a topic. In this configuration, the story and challenge that serve as the focal point of a PBL problem paint an image that can spark students' interest in a topic. For instance, the genetics strand (Chapter 7) includes stories about coat color in cats. Even if students are not familiar with dominant and recessive alleles, phenotypes, Punnett squares, or X and Y chromosomes, students may connect with questions about why cats' coat colors differ and how we can predict what colors a cat may inherit. When you use PBL as an introduction to a concept, the final solution may be delayed until after you have taught some concept-building lessons, including the ones you probably already use in your unit. The purpose of PBL in this case is to provoke a "need to know" response from students so they will understand why they need to learn about patterns of inheritance.

So PBL works as a summative assessment activity and as an introductory activity, but PBL has another important place in your curriculum. PBL problems can be used as concept-building activities. Many teachers do not think of the problems as a "main course" in the unit plan, often because they view PBL as an application activity that must follow introduction of vocabulary and concepts. But in our research and practice, we find that PBL can be an effective strategy for helping students learn new concepts.

The PBL problems build concepts in a different manner than direct teaching. Rather than *giving* information to students, a problem presents a reason for students to *find* or *construct* their own understanding. The analysis framework then helps scaffold the process of organizing their ideas and planning their search for new information. The final results are a product of students' pursuit of a solution to the problem that requires new understandings on their part.

This view of PBL as a concept builder also permits or even requires the teacher to incorporate experiments, investigations, and other experiences within the "research" phase of the PBL process. Assigning your existing lab activities, homework, and practice exercises might be the ideal strategy for addressing something in the "need to know" list. The difference is that *students* have identified the need for those activities rather the teacher prescribing a required activity. This is a powerful tool for motivating students to engage in your existing lessons.

But does PBL work as a way to help all students learn new ideas? Some critics have questioned whether a class of students with mixed prior knowledge all benefit from PBL. Can students with no experience with a concept succeed in a PBL lesson? Do students who have a strong understanding of the concepts still learn in this format? These questions are important considerations, and the PBL Project for Teachers collected data to answer these issues. Our research showed that even with a diverse group of learner needs, PBL helped 80.5% of learners improve their content knowledge (McConnell, Parker, and Eberhardt 2013b). These results were seen with teachers who began the lesson with little or no prior content knowledge and others who entered with a high level of content knowledge. Teachers in the project reported that PBL was effective in helping special needs students, high achievers, and the entire spectrum of learners in the classroom. For teachers who are expected to address specific content standards, these findings suggest that you can successfully build conceptual understanding through PBL.

Assigning Individual and Group Work

PBL lessons naturally lend themselves to group work. Researching the "need to know" list, sharing proposed solutions, and conducting investigations are all tasks that teachers usually plan to be carried out by groups of students. There are many grouping strategies that can be used in the PBL framework. Teachers should also consider how they will permit individuals to function in the PBL setting.

When your lessons reach a point when group work makes the most sense, you can use your normal procedures or policies about grouping. Our experience suggests that groups of four students seem to be most effective, although pairs can also function very well during the research phase. Groups of more than four may be unwieldy and may leave one or two students unoccupied with the tasks at hand. The general rule we recommend is to set group sizes at the number of students *needed* to complete the task. A lab may need groups of four or five. A simple online search may need groups of two or three.

There are also numerous ways to assign students to groups: student choice, randomly drawn names, deliberate teacher-selected groups based on ability levels (heterogeneous or homogeneous), or groups based on who works well together. The most common strategies in the literature suggest that teachers should assign groups rather than allowing students to self-select, but you need to choose the strategies that fit your teaching style and classroom culture.

An Alternative Use of Grouping

In the typical PBL lesson, as described in the vignette about Ms. Sampson's classroom in Chapter 3, the initial discussions of Page 1 and Page 2 of the story take place in a whole-class setting. The students read the story and generate a shared list of ideas in each of the three categories of the PBL analytical framework ("What do we know?" "What do we need to know?" "Hypotheses"). But there are opportunities to group students in this initial analysis of the problem, as well. One of the teachers in the PBL Project for Teachers, whom we will call Kylie, was very successful in creating and employing this strategy.

After her eighth-grade science class read Page 1 of a PBL story, Kylie would ask students to work in groups of three or four to discuss the story and generate their own lists under the three categories of the framework. Groups were responsible for recording their ideas in a PBL journal and were given a time limit. When the allotted time ended, Kylie would call on one member of each group and ask him or her to list one item from one of the categories. As Kylie recorded the ideas on the whiteboard, groups would continue adding ideas and discussing other groups' ideas until they had completed the first analysis. Then, the class would read Page 2 and repeat the group analysis once again.

After the students researched "need to know" topics, each group was asked to present a proposed solution that was recorded on the whiteboard, and the groups then discussed the list of proposals. The final discussion included a vote by groups on the best solutions and a discussion of the science concepts that relate to the problem and solutions. This strategy was very effective in keeping every student engaged in the initial analysis.

Individual Accountability

As in many science activities, one of the challenges you face is that it can be hard to know what each individual thinks or does during group work. In any science activity, a group of four may have two students actively doing an experiment while the other two observe … or find other distractions that take their attention away from the learning activity. In a whole-class discussion, it might be hard to include each student's ideas. PBL lessons present the same challenges.

Just as in other lessons, your knowledge about the individuals in your class can help you adjust your grouping to better engage the students in your classroom who might withdraw or take a more passive role in group activities. Consider giving individual students responsibility for finding information about one of the learning needs the group has identified, or consider using strategies to get each student to share at least one idea with the rest of the class. In the next section, we discuss strategies we have tested to help the teacher assess learning for each individual rather than relying only on a group presentation to evaluate every learner in the group.

Assessing Student Learning and Giving Feedback to Students

Early in the lesson planning process, teachers need to consider how they will assess learning. This is an important task for any lesson format, including PBL. The PBL Project for Teachers provided us with opportunities to develop, experiment with, and revise assessment strategies and instruments for PBL lessons (McConnell, Parker, and Eberhardt 2013a). The result is a multifaceted assessment plan that was applied to each problem in Chapters 5–8 of this book. In those chapters you will find open-ended general questions and application questions that can be used as pre- and post-assessments, transfer tasks to assess students' abilities to apply concepts, prompts for writing summaries of proposed solutions, and common beliefs inventories.

As part of our research and development, we tried several types of assessment questions. Some assessments were presented as multiple-choice "concept inventories" developed from national tests, but these did not give us much information about learners' ideas relating to *how* or *why* they chose their answers. We also tested concept maps as an assessment strategy, but these were difficult to evaluate because the learner was unable to describe the nature of the connections between ideas in detail. Interviews with individual learners can provide more insight into student thinking, but the time required to assess *every* student makes interviews impractical for most classroom settings.

The assessment questions that were most useful were open-response questions to which learners wrote their responses. Some teachers may find the responses to these open-ended questions very time-consuming to read, but no other type of question was effective in showing what ideas learners have and how they connect those ideas to each other. These connections were very revealing—they often exposed misconceptions and gaps in students' understanding.

In the rest of this chapter, we describe the various types of assessments and questions that may be used in a PBL lesson.

Pre- and Post-Assessments: General and Application Questions

As you teach a PBL lesson, it is important to assess what students know before you begin. For our research, we used a pre-assessment and an identical post-assessment to describe the new ideas learned and changes in teachers' ideas about science. You can do the same with your students. But developing questions that assess both the breadth and depth of your learners' ideas is complex.

In the PBL problems presented in Chapters 5–8, we have included the assessment questions we tested, revised, and retested. You will find two kinds of questions: *general questions* and *application questions*. Let's look at examples from the Pale Cats problem in Chapter 7, "Genetics Problems":

- *General question:* Explain what is meant by "DNA to protein to phenotype."
- *Application question:* Gregor Mendel, considered to be the founder of modern genetics, found that sometimes when he cross-pollinated (mated) two pea plants with red flowers, some of their offspring had white flowers. Explain these results in terms of the DNA, protein, and phenotype.

So why two types of questions? Aren't they asking the same thing? At first glance, they both appear to ask about the same concept, but the wording can lead to very different types of answers that reveal very different levels of understanding!

For the general question shown, our learners wrote brief explanations that reflected a description or "textbook definition." Sometimes the responses were a string of words with arrows to show connections, as shown in the following examples:

> *Model response:* A piece of DNA codes for specific proteins, and the protein is what gives the individual its phenotype.

> *Model response:* DNA → mRNA → chain of amino acids (protein) → phenotype

Both of these responses can be considered "correct," but what does it mean that a learner can write these? Does the learner understand the mechanisms, or has he or she memorized a definition from the book or the lecture notes? Would the learner be able to tell us what happens if a mistake in the process occurs? We believe it is impossible to tell from these responses. This "descriptive" knowledge is important, but we want students to know more than just a definition. We'd like them to understand *why* or *how* DNA controls phenotype and how it applies to traits such as coat color in cats or a disorder such as diabetes in humans. So we developed a second type of question.

In the application question for this flower-color question, we describe a specific example that demonstrates how a gene (piece of DNA) has variations that lead to different appearances (phenotype). The question then prompts the student to explain *how* or *why* this works, as shown in this example:

> *Model response:* There are two different versions of the gene—red and white. The red gene has instructions that tell the cell to make a red pigment protein. The white gene tells it to make a white pigment. If the flower has at least one red gene, the flower will be red. Otherwise, the flower will have only the white protein and will look white.

This response helps us see what the writer thinks about how DNA leads to a phenotype. Although this answer has some accurate information, it also reveals some inaccurate ideas. An accurate description would view the two variations as turning the pigment production either "on" or "off." The white "gene" actually fails to make a pigment protein for this trait, so white is the lack of pigment, not the presence of white pigment. The learner has also used the term *gene* inaccurately. There is only one gene (flower color), which has two variations (alleles)—red and white.

This application question requires the learner to "explain" rather than just describe or define. The written response to an application question is often longer than a response to a general question, and, as in this case, can show which *parts* of a concept the learner has mastered and which are not yet fully developed or accurate.

So why not just ask the application question? For many topics, learners tend to focus on a couple of key elements of an application question. The general question gives room for the student to add ideas that may be peripherally connected or to leave out concepts that are important. We found that if we first ask the general question, followed by one or two application questions, we gain insight into the overall awareness students have with a concept *and* how deeply they understand the idea. Together, these questions give us a great view of the writer's understanding of the role DNA plays in shaping our inherited traits.

The final sentence in the example application question (*Explain these results in terms of the DNA, protein, and phenotype.*) is also very important in getting a detailed explanation. In our first trials, we left that sentence out because we felt that this was leading the student to the answer. But without it, learners seemed to write incomplete answers. They might write about one important term and leave others out. In our later trials with the more specific prompt, we found that responses were more thorough but still included inaccurate or incomplete ideas and gaps in understanding. We also saw responses that overtly stated that the student did not know part of the response (e.g., *I don't know what the protein has to do with phenotype!*).

TEACHING TIP: Implementing the Two Types of Questions

Both the general questions and the application questions are a good way to assess what each *individual* student knows! We suggest you NOT use these as a group writing assignment.

In your classroom, you can use these questions as part of your own assessment plan. The questions are very well suited to be a pretest, either for a unit or for each specific PBL problem. These questions can also be used as post-assessment questions, either in the form of a quiz or open-write about the specific concept or as part of a chapter or unit test.

Along with each question, we have included a model response that you may use as a guide for developing a scoring key or a rubric for evaluating the responses.

STUDENT DRAWINGS

When you ask students to respond to the pre- and post-assessment questions, allow your students to include drawings to illustrate their ideas. In fact, encourage it! Their drawings often reveal misconceptions and accurate understandings better than text. Some learners are verbal, but others are visual, and their conceptual understandings may be easier to communicate in a drawing.

As an example, let's look at a drawing (Figure 4.2) that was a response to the following pretest question from the elementary life cycles strand (Chapter 5) in the PBL Project for Teachers:

> Look at this log. It weighs 10 pounds. Where did all the atoms that make up this log come from? Answer first in writing, and then answer using a labeled diagram. Both formats should stand as complete answers.

The pretest question refers to a concept addressed in the Baby, Baby Pear problem, which helps students learn that plants build their bodies largely from the carbon dioxide they take from the air.

Figure 4.2. Sample Student Drawing for Pretest Response

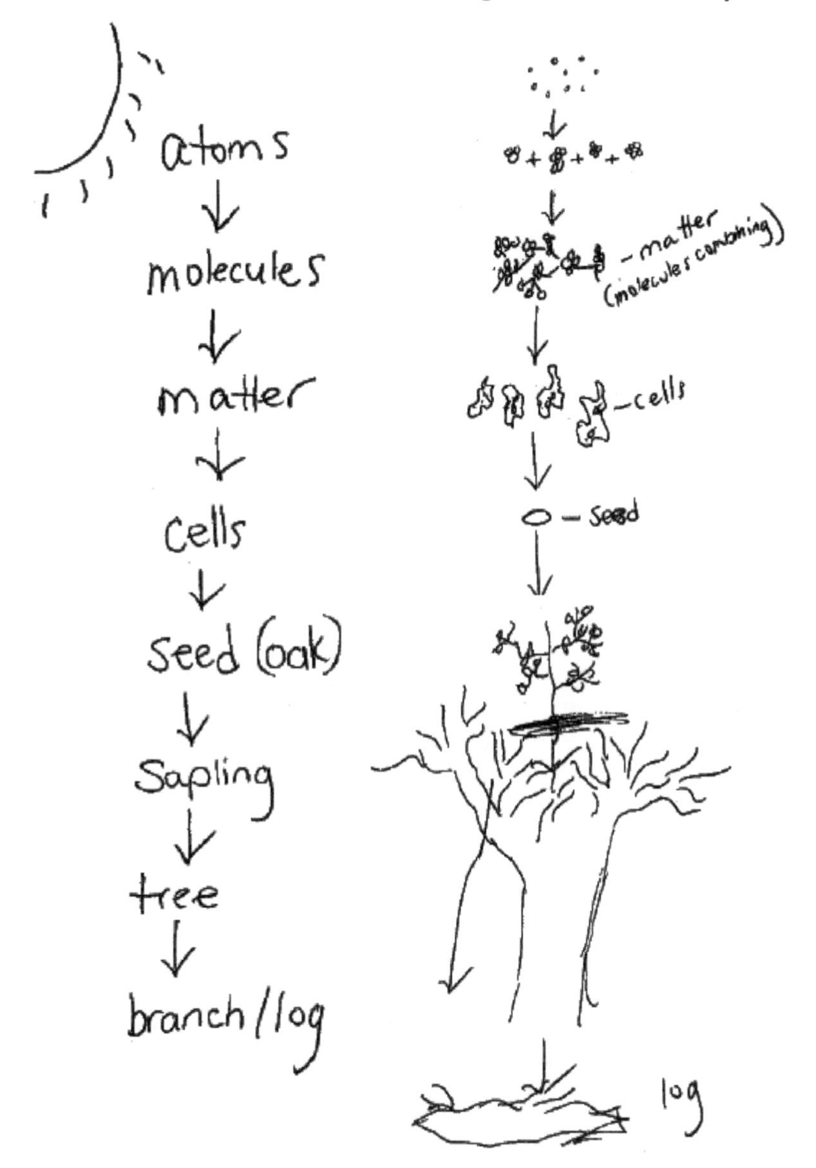

The text and diagram in Figure 4.2 suggest that the learner has a scientifically inaccurate view of how plants build the matter that makes up their tissues. Some indicators worth noticing are the idea that molecules join together to make cells and that cells join together to make seeds. Nothing in the response includes the concept of photosynthesis as a process carried out by cells to connect atoms together and build sugars that make up the wood (cell walls) in the tree. The response suggests that the learner is confusing levels of organization of matter and the process of building materials needed in the plant.

Transfer Tasks

Have you ever taught a concept, thought your students "got it," and then asked a question on a test or exam that represents the very same concept in a different context? Did your students fail to connect the two situations and recognize them as the same concept? All too often the answer to that question is "Yes!"

The problem of getting students to transfer their learning to new situations is nothing new (Ellis 1965). One of the things we want to know about PBL is how well students can transfer what they learn as they resolve a problem to new contexts. In each of the content chapters in this book, we have included *transfer tasks*—open-ended questions that ask students to apply the concept taught in a problem or group of problems to a new situation. Let's look at an example of a transfer task from the Baby, Baby Pear problem in Chapter 5, "Elementary Life Cycles Problems" (this is not the complete task):

> Many plants, such as carrots, can reproduce vegetatively as well as sexually. This means that rather than producing seeds, some part of the plant can generate all of the parts of a new plant. ... How is the life cycle of a carrot that undergoes vegetative reproduction similar to and different from that of the pear tree? What are some the benefits and costs of each life cycle?

In the Baby, Baby Pear problem, students explore the life cycle and development of a pear and compare them with animal development and life cycles. In the transfer task, students are asked to think about carrots that can grow from a section of root rather than just relying on seeds. To do so, students need to think about life cycles in the same way as in the story presented in the problem, but now they are applying the concept to a new and slightly different context. The model response included in the Assessment section of this problem goes into more detail than most students will write, but it gives some ideas about the language students could be expected to use in describing life cycles.

TEACHING TIP: Using the Transfer Tasks

Transfer tasks can be used in your lesson plan as a way to check understanding at the end of the PBL problem, either as a short stand-alone essay or as part of a chapter or unit test. This makes an excellent individual task to help you assess each student's learning.

Along with each transfer task, we have included a model response that you may use as a guide for developing a scoring key or a rubric for evaluating the responses.

Solution Summaries

Another important source of information about what your students have learned will be the solutions they develop as a final product of a PBL lesson. Individuals or groups build these solutions as an answer to the "challenge" presented on Page 1 of the problem. In some cases, the solutions may be brief, but in other problems, the solutions may be an extensive paper, a model, or even a demonstration. This type of assessment is flexible enough to be used as an individual assessment or a small-group assignment to help plan for discussions.

In Kylie's classroom, each group was required to share its solution and an explanation of why the group members thought their solution was correct. The class then voted for the solutions that made the most sense. This process was not a way to select the "right answer," but it supported Kylie's effort to foster discussion and evaluation of which solutions best matched the information collected. In many cases, more than one solution is correct, so a consensus agreement is not needed.

In our own use of PBL, this assessment usually entailed having individual learners write a page or two about their resolution of the problem. The benefit of an individual written solution is that it gives the teacher a way to assess the understanding of each student in the classroom. The individual assessments are important because much of the sharing of ideas throughout the analysis of the problem will have been done in groups. It is possible, even likely, that the teacher will hear many correct statements, but they come in bits and pieces from individual students. Some students may not have shared their ideas and questions, and others may have mastered one element of a complex problem or concept, but this "group speak" view of the understanding of the class as a whole does not necessarily translate to a complete understanding for each individual.

By asking each student to summarize the final solution to the problem, the teacher can find out which students have been able to connect concepts and apply them accurately and completely and which students still have gaps in their understanding, have misconceptions, or describe concepts in vague or "fuzzy" language. In some cases, the solutions will point to a need for the teacher to lead the class in additional activities, explain ideas to the whole class, or ask some probing questions in a discussion.

When the assessments used by the teacher are used to guide instructional decisions as described here, PBL becomes a rich context in which to frame assessment questions. The PBL lesson is not the only activity for teaching concepts in the unit, but it is a key component that we suggest gives the instructor a far better understanding of the depth and limits of what learners actually know and learn.

Let's look at an example from the "Where's Percho?" problem in Chapter 6, "Ecology Problems." In this problem, students investigate fishermen's concerns that a rising population of cormorants at Loon Lake is going to disrupt the perch fishing they enjoy. The following is a solution summary written by a learner in the PBL Project for Teachers:

> Nothing should be done at Loon Lake. The double-crested cormorants have not been linked to a decrease in Loon Lake's perch population. The cormorants are migratory birds that stop at Loon Lake for a perch dinner and continue to move on. There are not enough data to know how the birds affect

> the perch population in Loon Lake. The cormorants do not have very many predators and that could explain the increase in their population. Also, the decrease in pesticides increases the birth rate of the cormorants. It appears to be fishermen who don't like the cormorants because they decrease the chances of catching fish.

These solution summaries provide important evidence that teachers need as they teach a PBL lesson. They reveal learning outcomes, identify gaps in students' understandings, and help the teacher recognize the student's assimilation of ideas discussed in the group discussions or found in the research phase of the PBL process. For instance, the information about cormorants only being in the lake for a short time and the fact that decreasing pesticide levels have caused a higher birth rate in cormorants reveals some important learning. The writer also included a reference to the suggestion by fishermen that there must be a predator that can help control the cormorant population. But it also shows that the learner may not have learned that perch are only one species in the diverse diet of a cormorant, making the effect on the perch fishery less dramatic.

The information gaps or misconceptions in students' understanding gained from these written summaries of the solution are important tools in helping to assess learning and plan the next steps in instruction. But there are also tools available during the lesson that can provide even more data to help direct instructional decisions. These tools include the formative assessments discussed next.

Formative Assessments

The assessments described earlier are based on questions deliberately posed to learners as either a diagnostic or summative assessment. But like any teaching strategy, PBL presents many opportunities to conduct formative assessments of student thinking and learning. In fact, much of the structure of the PBL analytical discussion gives the teacher a clear picture of how students' ideas are changing. Formative assessments also allow the teacher to adjust, adapt, and revise plans on the fly. The following subsections describe some parts of the PBL process and some simple strategies teachers can add that act as a formative assessment.

PERFORMING AN INITIAL ANALYSIS

In earlier chapters, we suggested that the facilitator record ideas generated during the three-part analysis ("What do we know?" "What do we need to know?" "Hypotheses") on paper, a whiteboard, a SMART board, or some other form of display. The entire discussion of the story (Page 1 and Page 2) involves students sharing ideas, asking questions, and creating hypotheses. Each of these statements is a valuable bit of information for the teacher.

As you lead your class through a PBL lesson, it is important that your list of ideas be kept in some format you can store. Students nearly always change their ideas or revise hypotheses, and as they do this, you can see evidence of learning. By keeping each idea and either crossing out or modifying them, your permanent record helps you trace this development. Even your students will see and talk about how their ideas have changed.

We also have found that the lists help you identify opportunities to pause the PBL process long enough to answer a question or correct a misconception. You may also find that students' ideas and questions uncover a need to do certain hands-on investigations or demonstrations as part of the instructional process. Without a clear record of student-generated ideas, you may have a hard time catching those "teachable moments" or bringing your class back on track when the pause is finished.

REPORTING RESEARCH FINDINGS

After the initial analysis, the next opportunity to assess the development of students' ideas happens as the groups research ideas listed in the "need to know" list and share their findings with the class. Some teachers use graphic organizers to help students record and report information they find from text and online sources. These documents can be a source of information in your formative assessment plan.

But you can also implement your own plan to have students put their information on the board to share with the class or give a short presentation of their findings. As they share what they have found, you can collect and record their ideas to provide more evidence of learning or to identify emergent learning needs.

Strategies for keeping this information may include science notebooks or journals, a presentation paper or poster, electronic whiteboard presentations saved on the teacher's computer, or videotaped records of groups' presentations. The only limit on how the sharing of findings can be done is the imagination of teachers and students. We recommend you adapt the procedures you may already use when you ask students to share in front of the class.

ADMINISTERING EXIT TICKETS

PBL lessons are often a process that takes more than one class period, especially if you teach in 45- to 50-minute periods. A typical plan might include the initial analysis (Page 1 and Page 2) on the first day, followed by a class period to research information, and a third session for presentation and final discussion.

The end of each day presents another opportunity for formative assessment. Many of the facilitators in the PBL Project for Teachers used exit tickets (see description and vignette in Chapter 3, p. 42) as a way to get a brief look at what students are learning or what topics they are struggling to understand. Examples of exit tickets include statements that ask students to write reflections that give the teacher feedback about the lesson, such as "What solution do you think is most useful?" "What is one topic you are still confused about?" and "One thing you've learned is … " The facilitator or teacher can review these short anonymous assessments to make choices about what topics to address with students as the next class period begins or about how to help students understand difficult concepts.

Several books and websites offer ideas for these types of formative assessments. Examples of publications are *Seamless Assessment in Science* (Abell and Volkmann 2006), *Formative Assessment Strategies for Every Classroom* (Brookhart 2010), and "Formative Assessment Made Easy" (Cornelius 2013).

WRITING SUMMARIES OF THE "BIG IDEAS"

A variation of the exit ticket strategy that we tested in the PBL Project for Teachers was to ask learners to write a summary of the "Big Ideas" they learned from the PBL lesson. The Big Ideas are the foundational concepts that Roth and Garnier (2006) suggest teachers be mindful of when they plan and implement lessons. The lists of Big Ideas were generated by the participants in each content strand after they had reached the final resolution of a problem. Listing the Big Ideas they felt they had learned in the problem gave learners a chance to think metacognitively about the new ideas they were developing and focused their attention on the objective or learning goals of the lesson.

Our use of this assessment with teachers showed that sometimes we think learners are gaining knowledge of our target learning goals, but, in fact, they are thinking about other aspects of a problem. Because we want to know what the real outcomes are, these written summaries help us focus on the real changes in learners' thinking. Like the solution summaries described earlier, these summaries often point to opportunities for other instructional activities. Again, the assessment guides our next steps as we teach the concept.

TAKING COMMON BELIEFS INVENTORIES

A final form of assessment that we tested in many of the problems represented an attempt to see if PBL lessons could help correct inaccurate or unscientific understandings. As you have probably noticed, there are many science concepts for which people hold common beliefs, and some beliefs are not accurate. The common beliefs inventories were tested to see if they were able to help identify these misconceptions and if students' ideas about these concepts changed as they completed a PBL lesson. These assessments were not created for every problem, so we have included only a few examples in this book.

The common beliefs inventories were designed to be a brief measure to see what students think about common ideas and misconceptions. The inventories included about 10 true or false statements, and students were instructed to explain *why* each statement is true or false. This explanation is important because it probes more deeply into learners' ideas and avoids the trap of allowing an unexplained "guess."

The inventories proved helpful in some problems as a pre- or post-assessment. In the PBL Project for Teachers, we administered the common beliefs inventories at the beginning of the workshop and had participants revisit the questions after each problem was solved to see where new information might change their responses. The results showed instructors which common beliefs had changed and which remained consistent even after the lesson. In some strands, the inventories were given multiple times, and learners were asked to compare their own answers on each trial with previous tests. Letting students look back at their own learning is a powerful form of metacognition that help students become more reflective and self-regulating.

As an example of the information the teacher can gain from a common beliefs item, let's look at a genetics assessment. One of the true or false statements in the genetics Common Beliefs Inventory reads, "One gene makes one protein." This is a simple statement that is oversimplified; some genes act as a switch to "turn on" or "turn off" other genes, but when a gene codes for a protein, each gene codes for just one protein. Let's look at some responses to this statement listed by learners in the PBL Project for Teachers:

- **Laura:** True. One gene makes one protein, but a combination of genes can make a specific trait.
- **Amber:** False. One gene can make many proteins depending on the amino acid sequence.
- **Donna:** False. Many genes code for amino acids, which chain together to create proteins.
- **Tony:** False. Genes can make many proteins.
- **Janice:** True. The DNA can only be in one order, and it will have a "begin" and an "end" to denote the same order or amino acids.

Not only do the true or false answers vary, but the reasons behind their choices differ dramatically. Amber believes that a gene can make many proteins, but her response suggests she might be referring to different alleles of the same gene that make different versions of a protein. But Donna's answer reveals a possible misconception about what gene sequences code for when she says they "code for amino acids." The examples, models, and explanations a teacher chooses for a group of learners needs to take into account the specific ideas underlying the students' ideas, and the common beliefs inventories give yet another tool for assessing those ideas.

As you explore the various problems in Chapters 5–8, look for the different assessment questions and the model responses. You are welcome to use the assessments as we present them, but you may also wish to modify and adapt the questions.

References

Abell, S. K., and M. J. Volkmann. 2006. *Seamless assessment in science: A guide for elementary and middle school teachers.* Portsmouth, NH: Heinemann Educational Books.

Brookhart, S. M. 2010. *Formative assessment strategies for every classroom: An ASCD action tool.* Alexandria, VA: Association for Supervision and Curriculum Development.

Cornelius, K. E. 2013. Formative assessment made easy: Templates for collecting daily data in inclusive classrooms. *Teaching Exceptional Children* 45 (5): 14–21.

Ellis, H. C. 1965. *The transfer of learning.* London: Macmillan.

McConnell, T. J., J. M. Parker, and J. Eberhardt. 2013a. Assessing teachers' science content knowledge: A strategy for assessing depth of understanding. *Journal of Science Teacher Education* 24 (4): 717–743.

McConnell, T. J., J. M. Parker, and J. Eberhardt. 2013b. Problem-based learning as an effective strategy for science teacher professional development. *The Clearing House: A Journal of Educational Strategies, Issues and Ideas* 86 (6): 216–223.

Roth, K., and H. Garnier. 2006. What science teaching looks like: An international perspective. *Educational Leadership* 64 (4): 16–23.

ELEMENTARY LIFE CYCLES PROBLEMS

The problems in this chapter ask students to identify basic needs of living things and to describe simple life cycles of animals, plants, and fungi. The problems involve students in a number of science practices included in the *Next Generation Science Standards* (*NGSS*; NGSS Lead States 2013), most notably Developing and Using Models (science and engineering practice [SEP] 2) and Constructing Explanations (SEP 6). They also emphasize the crosscutting concept (CC) of Patterns—similarities in the needs and life cycles of living things.

Big Ideas

Life Cycles

- Despite the vast differences between the organisms, there are patterns in the life cycles of animals, plants, and fungi. All of these life cycles include being born or separating from a parent, developing into an adult, reproducing, and eventually dying. The details of the life cycles are different for different organisms.

Needs of Living Things

- There are patterns in basic needs of organisms. Animals, plants, and fungi need air, water, and food. Plants also require light.
- Animals, plants, and fungi use food as a source of energy and as a source of building material for growth and repair.
- Plants make their own food. Animals and fungi use materials from other living things as food.

Conceptual Barriers

As with most topics, one of the common problems is students seeing individual cases rather than patterns across cases. Given the variety of living things, it is not surprising that children see different organisms as unique. At first glance, kittens, cacti, and mold have different life cycles and needs.

Common Problems in Understanding

Teachers often get drawings from students with the following misrepresentations: (1) plants give off oxygen, and only animals take it in, and (2) only animals give off carbon dioxide.

Common Misconceptions

- Living organisms use food only as a source of energy.
- Plants do not need food as either a source of matter or energy.
- Fungi are plants.

Interdisciplinary Connections

There are a number of interdisciplinary connections that can be made with the problems in this chapter. For example, connections can be made to literature, language arts, geography, mathematics, and art and music (see examples in Box 5.1).

Box 5.1. Sample Interdisciplinary Connections for Elementary Life Cycles Problems

- **Literature:** Read one of several books about life cycles. Examples include *The Reason for a Flower: A Book About Flowers, Pollen, and Seeds* by Ruth Heller (1999), *Tadpole's Promise* by Jeanne Willis (2005), *The Very Hungry Caterpillar* by Eric Carle (1969), and *Like Jake and Me* by Mavis Jukes (1984).
- **Language arts:** Write a story or poem about the life cycle of a species, or produce a video story or a photo storyboard of a life cycle.
- **Geography:** Study the origins and habitat of hamsters in the wild, or track the migratory routes of monarch butterflies.
- **Mathematics:** Graph the time it takes different species to complete a life cycle, make timelines to scale of a species' life cycle, or measure the length and weight of the specimens observed.
- **Art and music:** Create a drawing, poster, sculpture, or song about life cycles. Modify the song "Over in the Meadow" (see the book by John Langstaff and Feodor Rojankovsky [1973]) to use the species in the life cycle examples.

References

Carle, E. 1969. *The very hungry caterpillar.* New York: Putnam.

Heller, R. 1999. *The reason for a flower: A book about flowers, pollen, and seeds.* New York: Penguin.

Jukes, M. 1984. *Like Jake and me*. New York: Random House.

Langstaff, J., and F. Rojankovsky. 1973. *Over in the meadow.* San Diego, CA: Harcourt Brace.

NGSS Lead States. 2013. *Next Generation Science Standards: For states, by states.* Washington, DC: National Academies Press. *www.nextgenscience.org/next-generation-science-standards.*

Willis, J. 2005. *Tadpole's promise*. London: Andersen Press.

Elementary Life Cycles Problem 1: Baby Hamster

Alignment With the *NGSS*

PERFORMANCE EXPECTATIONS	• *K-LS1-1:* Use observations to describe patterns of what plants and animals (including humans) need to survive. • *3-LS1-1:* Develop models to describe that organisms have unique and diverse life cycles but all have in common birth, growth, reproduction, and death.
SCIENCE AND ENGINEERING PRACTICES	• Planning and Carrying Out Investigations • Analyzing and Interpreting Data • Constructing Explanations and Designing Solutions
DISCIPLINARY CORE IDEAS	• *LS1.B-5 Growth and Development of Organisms:* Reproduction is essential to the continued existence of every kind of organism. Plants and animals have unique and diverse life cycles that include being born (sprouting in plants), growing, developing into adults, reproducing, and eventually dying. • *LS1.C-5 Organization for Matter and Energy Flow in Organisms:* Animals and plants alike generally need to take in air and water, animals must take in food, and plants need light and minerals; anaerobic life, such as bacteria in the gut, functions without air. Food provides animals with the materials they need for body repair and growth and is digested to release the energy they need to maintain body warmth and for motion. Plants acquire their material for growth chiefly from air and water and process matter they have formed to maintain their internal conditions (e.g., at night).
CROSSCUTTING CONCEPTS	• Patterns • Structure and Function

Keywords and Concepts

Mammalian life cycle, needs of living things

Problem Overview

Students observe the development of a classroom pet, a hamster, with a focus on patterns and comparison to other familiar animals and people. Full-color versions of all the problem's images are available on the book's Extras page at *www.nsta.org/pbl-lifescience*.

ELEMENTARY LIFE CYCLES PROBLEM 1

Page 1: The Story

Baby Hamster

One day, Charles came to school with a picture of baby hamsters. His hamster, Nibbler, had babies the night before. He wanted to know if anyone wanted to adopt a baby. Some of his classmates thought they were cute. Some thought they were ugly. But his classmates were all surprised at his hamsters' short legs, closed eyes, and lack of fur (see Figure 5.1). Their teacher, Ms. Grant, said the class could adopt one of the babies when it was old enough to leave its mother. In preparation, they had to figure out what their hamster would need and what its life cycle was so that they would be ready to care for it.

Your Challenge: *How do hamsters change as they grow, and what do they need as they change?*

Figure 5.1. Newborn Hamsters

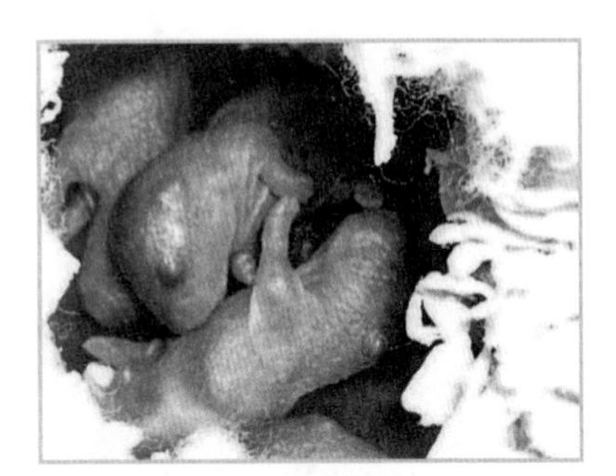

ELEMENTARY LIFE CYCLES PROBLEM 1

Page 2: More Information

Baby Hamster

- Hamsters (Figure 5.2) are members of an order of mammals called rodents. Chipmunks, mice, rats, squirrels, and beavers are also rodents.
- Like chipmunks, hamsters have extended cheek pouches for carrying food.
- Hamsters are nocturnal.
- Hamsters come from Syria, Greece, Romania, Belgium, and northern China.
- Hamsters may bite if they are handled by unfamiliar people.

Useful Links

- *www.livescience.com/27169-hamsters.html*
- *http://kids.nationalgeographic.com*

Your Challenge: *How do hamsters change as they grow, and what do they need as they change?*

Figure 5.2. Adult Hamster

ELEMENTARY LIFE CYCLES PROBLEM 1

Teacher Guide

Problem Context

This problem centers on hamsters because they are common classroom pets. However, any mammal will work well in this problem. It is useful if students can have some experience with live animals, even if it is just seeing a video.

The following response represents the ideas that are appropriate for students in grades 4 and above. Younger students should focus on the visible systems (the life cycle and basic needs) and not the internal systems.

> *Model response:* A hamster's life cycle has the same basic stages as those of all mammals, including humans [CC 1: Patterns]. The fertilized egg develops inside the mother, getting its food from her (through a placenta if not a marsupial). When the baby is first born, it gets its food from its mother in the form of milk. Once the baby is weaned, it must eat solid food and drink water. After some amount of growth and development, it is ready to have or contribute to making offspring of its own. At all stages, a hamster requires food, water, and oxygen. It uses food and water as sources of molecules for growth and repair. The food is also used with oxygen as an energy source.
>
> Hamsters also have many features of mammals and other animals that help them meet their needs and survive: legs for getting away from predators and getting to food, senses for monitoring their world, fur for warmth and protection, cardiovascular and digestive systems. These systems include heart, lungs, stomach, and intestines. Together, these systems take in food, water, and oxygen and get them to the cells throughout their bodies. At the same time, these systems collect and remove waste products from cells. Like other rodents, hamsters have teeth that grow throughout their lifetime. They use their teeth for many tasks, and the teeth would wear down if they weren't always growing.

Activity Guide

If students are unfamiliar with hamsters, there are many photos and videos available on sites such as YouTube.

Students can solve this problem with research but should be encouraged to draw on personal experiences with pets, animals at the zoo or pet store, animals in the wild, or animals seen on television. Rather than looking up facts, they can learn much by careful observation of live animals in the classroom or in videos (SEP 3: Planning and Carrying Out Investigations).

Safety Precautions

1. Remind students that animals in the wild may be diseased and could also attack or cause an injury. Always keep a safe distance between students and animals in the wild.

2. Remind students to treat animals in the classroom with respect and to handle them in a safe manner.

3. Instruct students to never touch the animal unless the teacher provides direction and permission.

4. Tell students to immediately report any animal bites to the teacher.

5. Research any restrictions on using animals during instruction. Some states or towns require veterinarian examinations for live animals in school classrooms.

6. Have students wash their hands with soap and water after handling animals in the lab.

ELEMENTARY LIFE CYCLES PROBLEM 1

Assessment

Baby Hamster

Transfer Task 1

Compare the life cycle of your favorite animal with that of a hamster.

- How are the life cycles alike and different?
- How do the stages of their life cycles compare?
- What body parts does each animal have that help it live during each stage?
- What does each animal need at each stage of life, and how do they use these materials?
- If you spent a day observing your favorite animal and a hamster as adults, what would you see them doing to fulfill their basic needs?

Transfer Task 2

Compare the life cycle of a hamster with that of a human (Figure 5.3).

- Use Table 5.1 (p. 74) to determine what percentage or fraction of their lives each animal spends in each stage. What do you think are the costs and benefits of the differences in the fraction of time spent in each stage?
- What body parts does each animal have that the other doesn't? What is the purpose of those body parts?

Figure 5.3. Different Stages of the Human Life Cycle

Table 5.1. Comparison of the Life Cycles of a Hamster and a Human

STAGE	TIME		FRACTION OF LIFE SPAN	
	Hamster	**Human**	**Hamster**	**Human**
Developing embryo	20 days	9 months	1.9%	1.0%
Nursing young	0–2 weeks	0–1 year	0.0–1.2%	0.0–1.3%
Eating solid food, protected by parent(s)	2–4 weeks	1–18 years	1.2%–2.5%	1.3%– 24.0%
Independent adult	4–36 weeks	18 years	2.5%	24.0%
Reproductive age	10 weeks	14 years	6.4%	19.0%
Total life span	3 years	75 years	100%	100%

Note: You can give students the chart with the percentages filled in or have students calculate the percentages.

Model response (for both transfer tasks): Hamsters and people have the same basic stages in their life cycles [CC 1: Patterns]. Hamsters reach adulthood and can have young much more quickly than people—in a matter of weeks rather than years—and they spend a smaller fraction of their lives growing up [SEP 4: Analyzing and Interpreting Data]. This means that they are ready to reproduce and pass on their genes sooner. However, people learn a lot before they have children, and this means they are more capable of caring for their young.

Hamsters have fur, continually growing teeth, and cheek pouches. The fur protects their skin and keeps in their body heat. The teeth and cheek pouches help them gather and eat tough or hard seeds. People have hands and a big brain and handle survival issues by learning and making tools [SEP 6: Constructing Explanations and Designing Solutions].

If I watched a giraffe (my favorite animal) and a hamster for a day, I would see both of them eat and drink, because they both need food and water [CC 1: Patterns]. In the wild, I think both of them would have to spend time looking for food and water. The giraffe would reach up high with its long neck and eat tree leaves. The hamster would poke around for seeds. When it found some, it would put them in its pouches and eat them later in its burrow, where the hamster would be safer.

Elementary Life Cycles Problem 2: Wogs and Wasps

Alignment With the *NGSS*

PERFORMANCE EXPECTATIONS	• *K-LS1-1:* Use observations to describe patterns of what plants and animals (including humans) need to survive. • *3-LS1-1:* Develop models to describe that organisms have unique and diverse life cycles but all have in common birth, growth, reproduction, and death.
SCIENCE AND ENGINEERING PRACTICES	• Planning and Carrying Out Investigations • Analyzing and Interpreting Data • Constructing Explanations and Designing Solutions
DISCIPLINARY CORE IDEAS	• *LS1.B-5 Growth and Development of Organisms:* Reproduction is essential to the continued existence of every kind of organism. Plants and animals have unique and diverse life cycles that include being born (sprouting in plants), growing, developing into adults, reproducing, and eventually dying. • *LS1.C-5 Organization for Matter and Energy Flow in Organisms:* Animals and plants alike generally need to take in air and water, animals must take in food, and plants need light and minerals; anaerobic life, such as bacteria in the gut, functions without air. Food provides animals with the materials they need for body repair and growth and is digested to release the energy they need to maintain body warmth and for motion. Plants acquire their material for growth chiefly from air and water and process matter they have formed to maintain their internal conditions (e.g., at night).
CROSSCUTTING CONCEPTS	• Patterns • Structure and Function

Keywords and Concepts

Insect and amphibian life cycles, needs of living things

Problem Overview

Students learn about the development of tadpoles and wasps and compare their life cycles with life cycles of mammals. Full-color versions of all the problem's images are available on the book's Extras page at *www.nsta.org/pbl-lifescience*.

ELEMENTARY LIFE CYCLES PROBLEM 2

Page 1: The Story

Wogs and Wasps

One day in April, Deja brought in some jelly-like eggs she found in the pond in the park (see Figure 5.4). She had them in a large water bottle, but Ms. Grant said they would be better off in the class aquarium. She thought that they were frog eggs. Because the class wasn't sure whether their goldfish (which had grown quite a lot over the school year) would eat them, they put the eggs in a net to prevent the fish from getting them. Ms. Grant said that the class had to figure out the frogs' life cycle and needs so that they could care for them properly.

Figure 5.4. Frog Eggs

The next day, Reganne brought in a picture of a "hive" she had seen hanging from a tree. The outside was papery and gray. The inside was a pale gray, but it had the hexagonal structure of a beehive. Some of the class wanted to adopt the "hive" also, but Ms. Grant identified it as a paper wasp nest (see Figure 5.5) that someone or something had torn open—not a safe classroom pet. However, Ms. Grant did say that part of the class could research the life cycles of bees and wasps while others were working on the frogs.

Your Challenge: *How do frogs, bees, and wasps change as they grow, and what are their needs as they change?*

Figure 5.5. Paper Wasp Nest

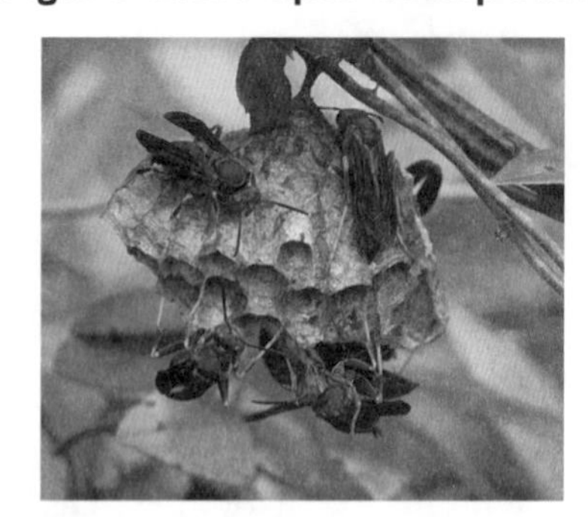

ELEMENTARY LIFE CYCLES PROBLEM 2

Page 2: More Information

Wogs and Wasps

The eggs that Deja found were from a northern leopard frog (see Figure 5.6). These frogs live throughout much of the northern United States, where ponds have vegetation along the shore and the water does not dry up for part of the year.

The wasps in Reganne's picture were indeed paper wasps. Like bees, paper wasps live in groups or colonies. They make their nests out of wood fibers that they chew up and mix with saliva. The saliva holds the fibers together. Unlike bees, wasps don't make honey or any other type of food to get them through the winter.

Your Challenge: *How do frogs, bees, and wasps change as they grow, and what are their needs as they change?*

Figure 5.6. Northern Leopard Frog

ELEMENTARY LIFE CYCLES PROBLEM 2

Teacher Guide

Wogs and Wasps

Problem Context

This problem focuses on animals that undergo metamorphosis—frogs, wasps, and bees. The problem can be modified to compare other insects that metamorphose (butterflies, moths, and mayflies, but not grasshoppers, which emerge from the egg looking very much like an adult).

The following responses represent the ideas that are appropriate for students in grades 4 and above. Younger students should focus on the visible systems (the life cycle and basic needs) and not the internal systems.

> *Model response (frogs):* Female frogs lay eggs, which male frogs fertilize. The developing frog embryo uses the egg as food. Once the tadpole (also called a pollywog) consumes the egg, it swims free. The pollywog has a tail, mouth, digestive tract, heart, and gills. It eats plants or insects and takes in oxygen through its gills. After several months, skin grows over the gills and legs develop, and the tail shrinks. Once the legs are big enough, the frog can move about on land and eat insects. The frog absorbs oxygen through its skin so it must keep its skin moist.

> *Model response (wasps and bees):* Wasps and bees begin life as fertilized eggs. The developing embryos use the egg as food. They hatch in a few days but have very little form. At this stage, they are called larvae and look like fat, short, white worms. The larvae are fed by workers and grow, shedding their skin as it gets too tight. After several cycles of growth and shedding, they emerge as pupae that look more like adults. They are still fed by workers and undergo another transformation to become adults. As adults, they find their own food and also feed undeveloped young. At this stage of the life cycle, they have three body segments: a head, a thorax with wings, and an abdomen with hearts. They have a digestive tract and take in oxygen through openings in each segment. Adult paper wasps eat soft insects such as caterpillars. Adult bees eat honey and pollen (a source of protein).

Activity Guide

This problem is set up to foster students' observation skills. Students should start by watching time-lapse videos (see Box 5.2) or observing live animals (SEP 3: Planning and Carrying Out Investigations). Mealworms are an inexpensive and easy-to-maintain metamorphic animal to keep in the classroom. Other possibilities are commercially available—sea monkeys (brine shrimp), painted lady butterfly eggs, hissing cockroaches, and frog eggs. Looking at dead animals such as wasps using a hand lens will allow students to observe specific body features associated with basic functions. Students can fill in or confirm facts with research.

Box 5.2. Time-Lapse Videos of Animals

- From frog eggs to tadpoles (pollywogs): *www.youtube.com/watch?v=B3d48VfQHbY*
- From tadpoles (pollywogs) to frogs: *www.youtube.com/watch?v=aTanXdFVEOw*
- From caterpillar to butterfly: *www.youtube.com/watch?v=J3w85HnC1HI; www.youtube.com/watch?v=m2VLkMoC274*
- From eggs to adult bees: *www.youtube.com/watch?v=f6mJ7e5YmnE*

To help students see patterns (CC 1: Patterns) across the life cycles of different organisms and understand the idea of cycles (organisms producing other organisms), they can draw the life cycles on paper plates. Have them start each life cycle with the newly fertilized organism at 12:00. 11:00 can be the death of the organism, with reproduction at 10:00. Students can fill in the intervening stages. They can draw what the organism looks like and indicate its requirements at each stage (SEP 2: Developing and Using Models; SEP 4: Analyzing and Interpreting Data).

Safety Precautions

1. Remind students to treat animals in the classroom with respect and to handle them in a safe manner.

2. Instruct students to never touch the animal unless the teacher provides direction and permission.

3. Tell students to immediately report any animal bites to the teacher.

4. Always check with the school nurse and parents for known allergies. Students may be allergic to certain insects (e.g., bees or wasps). You should know what to do as a first responder if a student with an allergy is stung by an insect.

5. Have students wash their hands with soap and water after handling any animals in the lab.

6. Research any restrictions on using animals during instruction. Some states or towns require veterinarian examinations for live animals in school classrooms.

ELEMENTARY LIFE CYCLES PROBLEM 2

Assessment

Wogs and Wasps

Transfer Task 1

You are an entomologist (a scientist who studies insects) and you have discovered a new type of beetle in the rain forests of Costa Rica. Because you are the first person to identify this new beetle, you get to name it. But before naming it, you want to study it so that you can give it a meaningful name. For your study, you plan to go to the place where you found the beetle and observe it every day for a month and write down what you see it doing. What are some important things you should look for, and what would these things tell you about the beetle? What information about the beetle might you not be able to find out about during your observation?

> *Model response:* I would observe what the beetle eats and how it moves around (Does it fly or walk?) and what it does if another (bigger) animal comes along. These observations would tell me what its basic needs (food and shelter/protection) are. If I am lucky and the timing is right, the beetle will mate with another beetle. That way, I can see how it reproduces and what earlier stages of the life cycle are like, but I might not see mating during my month of observations. Also, I don't know if this beetle is acting like other beetles of its kind. Maybe it just blew here, and this is not its usual habitat [SEP 3: Planning and Carrying Out Investigations].

Transfer Task 2

Butterflies lay eggs that hatch into tiny caterpillars. The caterpillars eat (usually the plant on which they were hatched) and grow (Figure 5.7, p. 82). When a caterpillar is big enough, it spins a chrysalis around itself. Inside the chrysalis, its body changes shape. When it comes out of the chrysalis, it is a butterfly with crumpled wings. The butterfly pumps fluid into the veins of its wings to straighten them out, and eventually it can fly away. As adults, many butterflies eat flower nectar.

How are the life cycles of butterflies, wasps, and frogs alike and different?

> *Model response:* All three organisms start as eggs. The body shape and form of what emerges from the eggs change/morph drastically before the organism becomes an adult able to reproduce. The organisms all need food during each stage of their lives.

The exception is the butterfly in the chrysalis. It uses the molecules in the caterpillar's body for an energy source and for new molecules for its transformed body.

The organisms have different body parts to help them survive. The butterflies and wasps fly away from predators, and frogs jump. Pollywogs and caterpillars are rather vulnerable to predators. Each of the organisms has a simple circulatory system and a digestive tract, but none of them have lungs for absorbing oxygen [CC 1: Patterns].

Figure 5.7. Caterpillar

Transfer Task 3

In the early stages of development, the embryos of mammals, birds, and fish are hard to tell apart. They look like the pollywog developing in its egg. A newly hatched fish most closely resembles the pollywog. It doesn't change much after it hatches. As you have just seen, pollywogs do change a lot as they develop into frogs. Bird and mammalian embryos also change a lot, but they do their changing before they are born. What are the benefits and costs of developing in an egg or parent versus developing after hatching?

Model response: For animals that develop after hatching, parents don't put much time or food into the developing young. Instead, they have lots and lots of babies to ensure that some survive. For animals that develop within a parent, the parents put a lot more time, food, and effort into each developing baby, but they cannot raise lots of babies in this labor-intensive way.

Elementary Life Cycles Problem 3: Humongous Fungus

Alignment With the *NGSS*

PERFORMANCE EXPECTATIONS	• *K-LS1-1:* Use observations to describe patterns of what plants and animals (including humans) need to survive. • *3-LS1-1:* Develop models to describe that organisms have unique and diverse life cycles but all have in common birth, growth, reproduction, and death.
SCIENCE AND ENGINEERING PRACTICES	• Planning and Carrying Out Investigations • Analyzing and Interpreting Data • Constructing Explanations and Designing Solutions
DISCIPLINARY CORE IDEAS	• *LS1.B-5 Growth and Development of Organisms:* Reproduction is essential to the continued existence of every kind of organism. Plants and animals have unique and diverse life cycles that include being born (sprouting in plants), growing, developing into adults, reproducing, and eventually dying. • *LS1.C-5 Organization for Matter and Energy Flow in Organisms:* Animals and plants alike generally need to take in air and water, animals must take in food, and plants need light and minerals; anaerobic life, such as bacteria in the gut, functions without air. Food provides animals with the materials they need for body repair and growth and is digested to release the energy they need to maintain body warmth and for motion. Plants acquire their material for growth chiefly from air and water and process matter they have formed to maintain their internal conditions (e.g., at night).
CROSSCUTTING CONCEPTS	• Patterns • Structure and Function

Keywords and Concepts

Fungus life cycle, needs of living things

Problem Overview

Students learn about the life cycle of a fungus and compare the life cycles and body structures of fungi with those of other types of living things. Full-color versions of all the problem's images are available on the book's Extras page at *www.nsta.org/pbl-lifescience*.

ELEMENTARY LIFE CYCLES PROBLEM 3

Page 1: The Story

Humongous Fungus

The students in Ms. Grant's class were thinking about record claims and how to test them. Ms. Grant was from Crystal Lake, Michigan, and her hometown wanted to get their humongous fungus into the *Guinness World Records* book, which is published annually.

Every summer, the people of Crystal Lake hold the Humongous Fungus Festival to celebrate the large fungus that grows nearby. It grows under 37 acres (an area bigger than 33 football fields) of forest in Iron County near the Wisconsin border. The fungus is believed to be more than 1,500 years old and to weigh about 100 tons—about the same as an adult blue whale. In the summer it produces golden mushrooms (see Figure 5.8), which are edible. The town is claiming that their fungus is the largest single fungus in the world that produces edible mushrooms.

Your Challenge: *Explain how an individual organism could get so big.*

Figure 5.8. Golden Mushroom

ELEMENTARY LIFE CYCLES PROBLEM 3

Page 2: More Information

Humongous Fungus

Ms. Grant's class thought that they could understand mushroom life cycles better if they learned about how to grow mushrooms. They found that people who raise mushrooms to eat do the following:

- Gather growth medium (material that the mushrooms will be grown on), which include leftovers from crops such as wheat straw, sawdust, manure, and compost.
- Kill all or most of the microbes in the growth medium by heating the material.
- Start the fungus growing by mixing in pieces of *mycelia*—the fine, rootlike structures of the fungus that grow throughout the "bedding" (Figure 5.9).
- Cover the mushroom beds with organic matter such as peat moss once the mycelia have filled out their growth medium. This step triggers the fungus to produce mushrooms, which in turn produce spores.

Figure 5.9. Fungus *Mycelia*

Note: A full-color version of this figure is available on the book's Extras page at *www.nsta.org/pbl-lifescience.*

The mushrooms produced by the Crystal Lake fungus are called honey mushrooms (*Armillaria gallica).* A related kind of fungus that grows in the Northwest is as big, if not bigger. This kind can kill the trees it grows under. Instead of consuming rotten wood like the Michigan fungus, this fungus may consume the roots of live trees. This fungus does not produce edible mushrooms.

Your Challenge: *Explain how an individual organism could get so big.*

ELEMENTARY LIFE CYCLES PROBLEM 3

Teacher Guide

Humongous Fungus

Problem Context

This problem focuses on the life cycle of mushrooms. However, it can be modified for any fungus, such as bread mold, fungi on a rotting log, or fungi on a compost pile.

> *Model response:* Mushrooms make spores, which, like seeds, can form new mushrooms. The mushroom is a special structure for producing and distributing spores. When spores land on the moist forest floor, they start to grow, producing long threadlike structures called *mycelia* that spread out through the soil. The mycelia secrete substances that digest the dead wood and plant matter in the soil and then absorb the molecules. This is their food, which they use as a source of molecules for growth and repair and as a source of energy along with oxygen. In the summer, the mycelia break the soil surface and form mushrooms.
>
> Under the right conditions, mycelia keep on growing, so if conditions remain stable for a long time (in this case, thousands of years), the fungus can grow very large. The stable conditions must include sufficient rain so that the ground doesn't dry out and so that new food is added each year (in this case, leaves falling every year). Mycelia are not able to transport nutrients very far, but the growth of the fungus is unlimited because it can simply grow more mycelia through the soil where its food source is found [CC 6: Structure and Function].

Activity Guide

Students can answer this problem mainly through careful observation of time-lapse videos, growing actual mushrooms (kits are commercially available), or examining mushrooms from the grocery store. Mycelia are often visible in compost piles (SEP 3: Planning and Carrying Out Investigations; SEP 6: Constructing Explanations and Designing Solutions).

Safety Precautions

1. Use only commercially available kits for growing mushrooms. Do not take mushrooms from the wild and grow them in the lab because they could be poisonous.

2. Never eat any plants brought into or used in the lab.

3. Wash hands with soap and water after completing the activity.

ELEMENTARY LIFE CYCLES PROBLEM 3

Assessment

Humongous Fungus

Transfer Task

In some ways, mushrooms resemble plants (Figure 5.10). They grow out of the ground and reproduce through spreading seeds or spores. In what way(s) are they different? How are they different from animals?

> *Model response:* Unlike plants, mushrooms or fungi do not make their own food. Mushrooms or fungi are like animals in that they need an external source of food. Unlike animals, however, mushrooms or fungi do not have a digestive tract for breaking down the food into smaller molecules that they can absorb. They secrete proteins that do the digesting outside of the fungus [CC 1: Patterns].

Figure 5.10. Mushrooms

Elementary Life Cycles Problem 4: Baby, Baby Pear

Alignment With the *NGSS*

PERFORMANCE EXPECTATIONS	• *K-LS1-1:* Use observations to describe patterns of what plants and animals (including humans) need to survive. • *3-LS1-1:* Develop models to describe that organisms have unique and diverse life cycles but all have in common birth, growth, reproduction, and death. • *5-LS1-1:* Support an argument that plants get the materials they need for growth chiefly from air and water.
SCIENCE AND ENGINEERING PRACTICES	• Planning and Carrying Out Investigations • Analyzing and Interpreting Data • Constructing Explanations and Designing Solutions
DISCIPLINARY CORE IDEAS	• *LS1.B-5 Growth and Development of Organisms:* Reproduction is essential to the continued existence of every kind of organism. Plants and animals have unique and diverse life cycles that include being born (sprouting in plants), growing, developing into adults, reproducing, and eventually dying. • *LS1.C-5 Organization for Matter and Energy Flow in Organisms:* Animals and plants alike generally need to take in air and water, animals must take in food, and plants need light and minerals; anaerobic life, such as bacteria in the gut, functions without air. Food provides animals with the materials they need for body repair and growth and is digested to release the energy they need to maintain body warmth and for motion. Plants acquire their material for growth chiefly from air and water and process matter they have formed to maintain their internal conditions (e.g., at night).
CROSSCUTTING CONCEPTS	• Patterns • Structure and Function

Keywords and Concepts

Plant life cycles, needs of living things

Problem Overview

A child wonders about how a pear grows on a pear tree and learns about the life cycle of a plant. (*Note:* The story in this problem is presented in two parts. We recommend that you present both parts of Page 1 before presenting Page 2.) Full-color versions of all the problem's images are available on the book's Extras page at *www.nsta.org/pbl-lifescience.*

ELEMENTARY LIFE CYCLES PROBLEM 4

Page 1: The Story—Part 1

Baby, Baby Pear

Sahar's mother liked to tell the story of Sahar's birth. She always started the story the same way:

> My husband is a biology nerd! How many people have an animal skull collection in their china cabinet and a microscope on the kitchen counter with insect eggs on a slide? He was so excited 12 years ago when I told him we were going to have a baby that he actually started a biology project in honor of the occasion! Our pear tree was in blossom at the time (Figure 5.11), and he watched it until he saw a bee moving from blossom to blossom. Then he chose one blossom that had its inner parts covered with pollen. He covered that blossom and its twig with a clear glass bottle. Then, we observed what happened over the course of the summer.
>
> Over the next six months, the pear blossom changed before our eyes. The petals fell off, and the ovary at the base of the blossom grew larger and began to look like a baby pear, only green in color. Pretty soon the growing pear didn't look anything like a flower. By September, the pear fruit almost filled the glass bottle and had turned a golden yellow.
>
> One night, we decided to liberate the pear from its glass cage. Once freed, my husband cut into the pear to count all the seeds inside. Then we enjoyed the sweet, juicy fruit. That night, I went into labor and the next day delivered a 7-pound baby girl—Sahar!

Figure 5.11. Pear Blossom

Your Challenge: *How did a pear blossom turn into a pear and then a tree?*

ELEMENTARY LIFE CYCLES PROBLEM 4

Page 1: The Story—Part 2

Baby, Baby Pear

Sahar's father started a new experiment when his daughter was born. First, he weighed one of the pear seeds. Then, he filled a huge clear glass container with soil and weighed the filled container. Finally, he planted the pear seed in it. Then he made two growth charts—one for the pear tree, one for baby Sahar—which he posted on the refrigerator door. Every time the baby had a checkup with the doctor, he wrote down the baby's weight and made some notes about what she was doing. On the same day, he checked the pear tree and posted its weight (always subtracting out the weight of the soil and the container) and noted how it had changed. Every year on Sahar's birthday, he added 3 ounces of fertilizer to the pear tree's water.

Figure 5.12. Pear Tree

Sahar and the pear tree grew up together. When the pear tree was 4 years old, it bloomed and grew three pears. When Sahar was 4 years old, she started preschool. When the pear tree was 9 years old, it was infected with fire blight. All the leaves turned black and looked like they had been burned. The next year, the tree produced lots of pears, but when Sahar and her parents bit into them they saw that they were filled with the larvae of the codling moth! That was a bad year for Sahar, too. She got the flu in February and broke her arm falling out of the pear tree in April.

Over the years, the pear tree provided habitat for nesting robins, tree frogs, and raccoons. There was often a large toad sitting under the tree, and Sahar was usually somewhere nearby.

Finally, when Sahar turned 13 and became a teenager, her father ended the project. (Figure 5.12 shows how the fully grown pear tree would have looked.) He weighed the tree in its container one last time and then broke the glass container. He removed all the soil from the roots as best he could, washed the remaining soil off the roots, and let the whole tree dry out completely. This took months! Then he weighed it one last time, calling this the *dry weight*.

Your Challenge: *How did a pear blossom turn into a pear and then a tree?*

ELEMENTARY LIFE CYCLES PROBLEM 4

Page 2: More Information

Baby, Baby Pear

The following chart shows sample data from Sahar's father's experiment.

Developmental Milestones of Sahar and the Pear Tree

AGE	MASS OF CHILD	DEVELOPMENTAL MILESTONES OF CHILD	MASS OF PEAR TREE	DEVELOPMENTAL MILESTONES OF PEAR TREE
0.5 years	16 lb	All parts present	0.005 lb	Thin stem and two leaves
1 year	21 lb	More hair; walking and babbling	1 lb	Roots, stem, and leaves
2 years	27 lb	Starting to say words	3 lb	Bigger stem and leaves
5 years	41 lb	Not growing as fast; can read	42 lb	Woody trunk, flowers, and few fruits
10 years	72 lb	Taller and heavier; arm out of cast	108 lb	Thicker trunk, lots of flowers, and fruit but full of moth larvae
13 years	101 lb	Going through a growth spurt	197 lb 110 lb, dry weight	Thicker trunk, lots of flowers, fruit small, and sweet

Your Challenge: *How did a pear blossom turn into a pear and then a tree?*

ELEMENTARY LIFE CYCLES PROBLEM 4

Teacher Guide

Baby, Baby Pear

Problem Context

This problem contrasts the life cycle of humans with that of trees. However, any plant can be used in place of the pear tree. If students have experience growing plants, that plant could be substituted or added to the story. Alternatively, any plant whose life cycle is captured by time-lapse photography works well.

Model response: The pear tree that "grew up" with Sahar started as a flower blossom. A bee with pollen on its legs visited the blossom, and some of the pollen was wiped on to the part of the ovary that sticks out. In this way the pollen grains fertilized a number of eggs in the ovary, triggering the ovary to grow into a pear. The fertilized eggs developed into seeds inside the pear. The fruit and the seeds used food made by the tree as a source of molecules they needed and as a source of energy.

The seed that Sahar's father planted contained the fertilized egg as well as food for the new plant. The seedling used the food in the seed as a source of molecules it needed to grow a stem, root, and leaves and as a source of energy. Once the seedling had leaves, it no longer depended on the food of the seed. It could make its own food. It used carbon dioxide from the air, water, and sunlight to makes its food. The growing tree used the food as a source of energy and, along with a small amount of minerals from the soil, built the food into the molecules it needed to grow.

Flowers have special parts that are important for their role in reproduction. The ovary contains special cells waiting to be fertilized. The pollen contains the other kinds of cells that participate in fertilization. The flower holds the pollen where insects will rub against it when they come to drink some of the sweet nectar and then carry it to other plants where it can fertilize eggs. Other types of plants hold the pollen where the wind will catch it and carry it to other plants.

Plants and trees have parts that help them make and use food. The roots take in water they need to make food (and move food through their bodies). The roots also take up small amounts of minerals that plants combine with atoms from food to make the molecules they need for growth. The leaves are the food-making systems. The stem or trunk holds them up so that they can catch light for food making. The leaves are thin so that the gases they need from the air can get in and out.

Activity Guide

Students can solve this problem and answer their questions by finding and reading the appropriate material. However, we believe it is important to help students visualize the tree's life cycle. Then they can think about and do investigations about where the tree is getting food at each stage and how the visible parts of the plant contribute to this (SEP 3: Planning and Carrying Out Investigations; SEP 6: Constructing Explanations and Designing Solutions).

Watching videos (e.g., a time-lapse video of a tree growing from a seed or a video of a bee pollinating a fruit tree) can help students visualize the tree's life cycle. Students can also do investigations of each stage of the life cycle:

- Germination—germinating seeds (corn, peas, grass, radish) on wet paper towels in baggies
- Growth and development
  - o Putting plants (e.g., celery, especially with leaves, or Queen Anne's lace) in water with food coloring (the colored water stains the capillaries in the stem) allows students to see how water moves up through the plant.
  - o Growing plants hydroponically (grass seed on sponges works well) helps students realize that plants are not using soil as food. The plants will grow well if their water is supplemented with fertilizer, but they will grow a reasonable amount in tap water.
- Flowering—students can dissect a flower and then gently gather pollen and rub it on the female part of the plant
- Students can grow and watch the whole life cycle of plants using Wisconsin Fast Plants over the span of about a month (*www.fastplants.org*).

Teachers can use the book *The Reason for a Flower* by Ruth Heller (1999) to connect to this problem. Additionally, the Annenberg Foundation provides a series of videos for teachers with clips of students discussing various topics. The clips can be used as sources of fruitful discussion questions or as sources of ideas to which students can react. The video course, Essential Science for Teachers: Life Science, is available at *www.learner.org/channel/series179.html* and includes the following sessions:

- "Session 1. What Is Life?": What distinguishes living things from dead and nonliving things? No single characteristic is enough to define what is meant by "life." In this session, five characteristics are introduced as unifying themes in the living world.
- "Session 2. Classifying Living Things": How can we make sense of the living world? During this session, a systematic approach to biological classification is introduced as a starting point for understanding the nature of the remarkable diversity of life on Earth.
- "Session 3. Animal Life Cycles": One characteristic of all life forms is a life cycle—from reproduction in one generation to reproduction in the next. This session introduces life cycles by focusing on continuity of life in the Animal Kingdom. In addition to considering what aspects of life cycles can be observed directly, the underlying role of DNA as the hereditary material is explored.
- "Session 4. Plant Life Cycles": What is a plant? One distinguishing feature of members of the Plant Kingdom is their life cycle. In this session, flowering plants serve as examples for studying the plant life cycle by considering the roles of seeds, flowers, and fruits. A comparison to animal life cycles reveals some surprising similarities and intriguing differences.

Safety Precautions

1. If students use scalpels for dissection, remind them that scalpels are sharp and can cut skin. Demonstrate the proper technique for using a scalpel.

2. Use caution when handling seeds for germination. They may contain pesticides.

3. When growing plants, use commercially available, sterilized soil that is free of pesticides and herbicides.

4. Wash hands with soap and water after working with seeds, plants, or soil.

ELEMENTARY LIFE CYCLES PROBLEM 4

Assessment

Baby, Baby Pear

Transfer Task 1

Sunflowers are *annuals*—plants that grow, reproduce, and die all in one year or growing season (Figure 5.13). How is the life cycle of a sunflower similar to and different from that of a pear tree? What are some of the benefits and costs of each life cycle?

Model response: The sunflower goes through all of the same stages as the pear tree: a seed sprouts a root and a stem with leaflets [CC 1: Patterns]. Once the leaves have a big enough surface exposed to the Sun, they undergo photosynthesis. The plant uses the food it makes through this process as a source of molecules to build more plant material and as an energy source. Once the plant is big enough, it flowers. When the flower is pollinated, the plant grows a fruit or seed around it. This provides food for the new plant to grow until it can do its own photosynthesis.

Figure 5.13. Sunflowers

The life cycle of the sunflower is much shorter than that of a pear tree—one year versus many years. Another difference is that the pear tree produces fruit and seeds for multiple years, but the sunflower produces its seeds all at one time. To produce seeds every year, the tree has to have a way of storing its food from one year to the next. Because the tree will probably live for many years, it can afford to build some of its precious food into reusable structures such as the trunk and branches.

Transfer Task 2

Many plants, such as carrots, can reproduce vegetatively as well as sexually. This means that rather than producing seeds, some part of the plant can generate all of the parts of a new plant. For this task, you will grow a new carrot plant from the top of a carrot.

Put about an inch-long section from the top of the carrot (with or without leaf stem) in a dish with enough water to cover the lower quarter to half of the carrot piece. Let it sit for four to seven days in a warm sunny area; you should see leaves developing. Observe the plant every day and note any changes. Once the leaves have developed fully, you can regenerate the orange root by transplanting the carrot to some loose, sandy soil. Continue to observe the top of your plant. Once a week, you can gently scrape back the soil to see how the root is developing.

How is the life cycle of a carrot that undergoes vegetative reproduction similar to and different from that of the pear tree? What are some of the benefits and costs of each life cycle?

Model response: The stages of vegetative reproduction for a carrot are as follows:

1. A plant part (root or part of the root and leaves) forms new rootlets and leaves.

2. Eventually, the new carrot root grows a full head of leaves, using energy and molecules from the large orange root.

3. When it has enough leaves, the carrot begins to provide itself with food from photosynthesis.

This is a faster way to reproduce, but the new carrot plant produced vegetatively will be the same genetically as the plant it came from [SEP 3: Planning and Carrying Out Investigations; CC 1: Patterns].

Elementary Life Cycles Problems: General Assessment

General Question

We have looked at a variety of living things: hamsters, frogs, wasps, mushrooms, pear trees, and sunflowers. In what way(s) are the life cycles of all of these organisms alike? In what way(s) do these organisims all have similar needs?

> *Model response:* All these living things come from another living thing (a parent). They grow and change, produce new individuals, and die. The growth, change, and reproduction require food as a source of molecules and energy [CC 1: Patterns].

Application Question

Think about [teacher chooses or has students choose two organisms from the list in the previous paragraph]. They don't have many features in common. If you were to observe them living in the wild, you would see them doing different things. But sometimes those different things mean that they have similar life cycles and needs. Describe something that you might see each organism doing that means that both have a similar life cycle or needs.

(*Note to teacher:* We recommend choosing two plants or two animals for younger students. Older students can compare a plant with an animal. This is a difficult task that might warrant working an example as a whole class. If students have made observations of live animals, encourage them to use their observation notes.)

> *Model response (for comparison of frog and pear tree):* The tadpole came out of a jelly-like egg. The pear tree grew from a seed. Both the egg and the seed came from a parent, which shows how both frogs and pear trees come from their parents. We saw the tadpole eating algae from the side of the fishbowl. Trees don't eat, but they have to have air, water, and sunlight to grow. Trees make their own food, but both the tadpole and the tree need food [SEP 7: Engaging in Argument From Evidence; CC 1: Patterns].

Common Beliefs

Indicate whether the following statements are true (T) or false (F), and explain why you think so *(model responses shown in italics).*

1. All the food you eat is either burned immediately to release energy or is stored to be used later as an energy source. *(F) Some of the food is used as a source of atoms and molecules that your body needs for growing and repairing.*

2. In organisms, food is stored in the body in the same form it is used for growth and repair. *(F) Different organisms store food in different molecules—oils and starches in plants, fat in animals. When they use the stored food to build, they change the stored molecules into other molecules that they need for growth and repair through a series of chemical reactions.*

ECOLOGY PROBLEMS

The ecology problems in this chapter lead the learner to examine the natural and human-made factors that affect diversity in local ecosystems. The problems address both descriptions and explanations of observations regarding food webs, ecological succession, the influence of invasive species, interspecific competition, and the choices humans make related to these topics. Descriptions of the observations are part of the elementary and middle school standards (grades 5–8), whereas high school students are expected to learn varying levels of detail about the interactions between species in ecosystems.

The ecology problems focus on the *Next Generation Science Standards* (*NGSS*; NGSS Lead States 2013) crosscutting concepts (CCs) of Patterns, Cause and Effect: Mechanism and Explanation, and Systems and System Models. The activities ask learners to consider how ecosystems respond to changing factors and how we can build models of an ecosystem to help us understand the patterns we see in nature. The solutions to the problems require learners to find the causal relationships between events taking place in our environment.

The ecology problems involve students in a number of science practices included in the *NGSS*, most notably Developing and Using Models (science and engineering practice [SEP] 2), Analyzing and Interpreting Data (SEP 4), and Obtaining, Evaluating, and Communicating Information (SEP 8). In addition, the problem-based learning (PBL) process involves students in Asking Questions and Defining Problems (SEP 1) and Engaging in Argument From Evidence (SEP 7).

Big Ideas

Interdependent Relationships in Ecosystems

- Organisms fill specific trophic levels in an ecosystem.
- An organism's niche is determined by the physical characteristics of the environment, the availability of food, and the presence of other species, including predators, prey, and competitors.
- If two species in the same ecosystem have similar niches, they will compete for limited resources. When one of the species has a competitive advantage, the other will decline in number or disappear from the ecosystem.

- The carrying capacity for a species depends on the availability of food in the trophic level on which it feeds (10% rule).
- If a prey species (e.g., alewife) declines, the biomass of predators (e.g., salmon) that can survive in the ecosystem also declines.
- The effect of the introduction of a new predatory species into an ecosystem depends on the variety of prey eaten by the predator and the degree of interconnections in the existing food web.
- Factors that increase decomposition of wastes and organic matter cause an increase in bacteria that consume oxygen, reducing the dissolved oxygen (DO) levels.

Ecological Succession

- Ecosystems change over time (ecological succession).
- Succession occurs when the growth of some species (e.g., sphagnum moss) leads to changes in the environment (e.g., buildup of sediment) that affect other species (e.g., shore plants).
- Glacial lakes form bogs when lack of water movement allows sphagnum moss to grow and create an acidic environment.
- As a bog ages, conditions at the edges change to allow different plants to grow and spread into the bog.
- After many years, the bog will fill in, become less acidic, and have more organic material and nitrogen available for plants. The ecosystem will become a beech-maple forest.

Natural Resources

- Each species has specific requirements for food, water, shelter, and (for aquatic species) DO.
- Some species cannot survive in lakes with low DO levels.
- DO levels can vary naturally with seasons or because of changes in conditions related to organic nutrients and decomposition.
- Changes in amounts of decomposition can be naturally occurring (eutrophication) or caused by human activities.
- Organisms need certain resources to survive in their environment.

Human Impacts on Earth Systems

- Invasive species may share a similar niche with native species. They sometimes will have a competitive advantage and can become a dominant species in their new environment.
- Introduction of an invasive species (e.g., zebra mussel) can create competition for a resource (e.g., plankton), reducing the number of native organisms in the ecosystem (e.g., alewife). In some cases, invasive species will feed on native species (e.g., round goby eat bass eggs).
- When managing populations, biologists need to consider the food source of the managed species and the entire food web that supports that species.
- DO levels can vary naturally with seasons or because of changes in conditions related to organic nutrients and decomposition.
- Changes in amounts of decomposition can be naturally occurring (eutrophication) or caused by human activities. Examples include fertilizer runoff and fecal matter from farms and septic systems.

Conceptual Barriers

Common Problems in Understanding

The findings of the research conducted in the PBL Project for Teachers revealed that many people are familiar with specific topics related to threats to biodiversity. However, individuals often lack an ecological understanding of the relationships between events and their causes. For instance, most people recognize that invasive species are harmful, but they do not understand that competition with native species is the issue rather than invasive species killing native species. Most people also believe that any loss of diversity in their local ecosystem can be attributed to human impacts. They do not know about natural changes in diversity such as ecological succession that will eventually limit diversity without human interference.

Common Misconceptions

The problems in this chapter address common misconceptions about invasive species, water pollution, and competition (see the subsections that follow). A pattern in these misconceptions suggests that most derive from partial understandings of concepts. These alternative understandings are associated with single issues that are frequently seen in newspaper, television, and internet stories about local issues. The misconceptions also show a lack of connection between concepts. The CCs in the *NGSS* are intended to focus attention on the overlapping nature of science concepts. In ecology, these crosscutting ideas are critical in applying new concepts. The implication for teachers is that students not only need to know about a single concept but also need to see the relationships between scientific concepts, preferably in the context of real-world situations. This will help students recognize the interrelatedness of ideas and the complexity of ecological problems.

COMMON MISCONCEPTIONS ABOUT INVASIVE SPECIES

- Changes in one population affect the populations of only those organisms that feed on it or are eaten by it.
- Invasive species are introduced only by humans.
- All invasive species are better at surviving than the native species.
- Invasive species "attack" native species and adapt faster than native species.
- Invasive species "consume" or "kill off" native species.
- Invasive species have a broader niche that is not as specific to a given habitat.
- Invasive species affect only the species with which they compete.

COMMON MISCONCEPTIONS ABOUT WATER POLLUTION

- When species that require high water quality disappear from a lake, it indicates some form of chemical pollution.
- Humans are the only source of pollution or cause of decline in water quality.
- All lakes in an area will have the same species, so any pollutant will affect all lakes in the same way.

COMMON MISCONCEPTIONS ABOUT COMPETITION

- Changes in one population affect the populations of only those organisms that feed on it or are eaten by it.
- Changes in an ecosystem are a sign of something "wrong" with the environment and are probably caused by humans.
- Only animals compete for resources.
- Competition involves the "stronger" species killing or fighting off weaker ones.

Interdisciplinary Connections

A number of interdisciplinary connections can be made with the problems in this chapter. For example, connections can be made to language arts, geography, social studies, mathematics, and art and music (see Box 6.1).

Box 6.1. Sample Interdisciplinary Connections for Ecology Problems

- **Language arts:** Write a letter to the local board creating resource management policies, or write a letter to the editor. Write a persuasive letter to wildlife managers to help propose new management strategies. Create an infographic about cormorants' role in the Great Lakes ecosystems.
- **Geography:** Use online mapping tools to find the location of one of the problems on a map, and find a similar location near the school that may have a related issue. Use geographic information systems software to map and compare ranges of invasive and native species.
- **Social studies:** Investigate the role of state and local governments in regulating use of natural resources and in regulating other environmental issues related to the problems in this chapter. Compare a Great Lakes salmon fishery to a Pacific Coast fishery where salmon are a native species, and look for similar features.
- **Mathematics:** Determine the number of fish a cormorant eats and extrapolate to find the effect of a flock of cormorants; compare fish consumption by cormorants to data on fishery use by humans. Build a graphical representation of the biomass of salmon, the small fish they feed on, and the plankton that make up the base of the food chain.
- **Art and music:** Create a collage, painting, or photographic work featuring the interrelated species in an ecosystem affected by invasive species. Write a song, poem, or creative story about interactions within an ecosystem or food web. Create a visitor's brochure as a trail guide to a local park, wetland, or bog that explains how invasive species affect the ecosystem.

Reference

NGSS Lead States. 2013. *Next Generation Science Standards: For states, by states.* Washington, DC: National Academies Press. *www.nextgenscience.org/next-generation-science-standards.*

Ecology Problem 1: Where's Percho?

Alignment With the *NGSS*

PERFORMANCE EXPECTATIONS	• *MS-LS2-1:* Analyze and interpret data to provide evidence for the effects of resource availability on organisms and populations of organisms in an ecosystem. • *MS-LS2-2:* Construct an explanation that predicts patterns of interactions among organisms across multiple ecosystems. • *MS-ESS3-3:* Apply scientific principles to design a method for monitoring and minimizing a human impact on the environment. • *MS-ESS3-4:* Construct an argument supported by evidence for how increases in human population and per-capita consumption of natural resources impact Earth's systems.
SCIENCE AND ENGINEERING PRACTICES	• Asking questions and Defining Problems • Developing and Using Models • Constructing Explanations and Designing Solutions • Engaging in Argument From Evidence • Obtaining, Evaluating, and Communicating Information
DISCIPLINARY CORE IDEAS	• *LS2.A-MS Interdependent Relationships in Ecosystems:* Predatory interactions may reduce the number of organisms or eliminate whole populations of organisms. Mutually beneficial interactions, in contrast, may become so interdependent that each organism requires the other for survival. Although the species involved in these competitive, predatory, and mutually beneficial interactions vary across ecosystems, the patterns of interactions of organisms with their environments, both living and nonliving, are shared. • *LS2.C-MS Ecosystem Dynamics, Functioning, and Resilience:* Ecosystems are dynamic in nature; their characteristics can vary over time. Disruptions to any physical or biological component of an ecosystem can lead to shifts in all its populations.
CROSSCUTTING CONCEPTS	• Cause and Effect: Mechanism and Explanation • Systems and System Models

Keywords and Concepts

Competition, food webs

Problem Overview

Fishermen in the Great Lakes are concerned that a growing population of cormorants will harm the perch fishery. Students are asked to evaluate the problem and propose possible actions. Full-color versions of all the problem's images are available on the book's Extras page at *www.nsta.org/pbl-lifescience.*

ECOLOGY PROBLEM 1

Page 1: The Story

Where's Percho?

In May, a flock of 27 double-crested cormorants (Figure 6.1) roosted and fed on Loon Lake, which is located in Montcalm County, Michigan. The residents surrounding the lake were concerned about the predatory birds feeding on the fish. They had heard of a similar problem at Les Cheneaux Islands. Fishermen in that area noticed a marked decline in the yellow perch (Figure 6.2) catch and attributed it to a high population of cormorants preying specifically on perch.

Your Challenge: *Determine if the fishermen of Loon Lake should be concerned with the presence of the fish-eating birds. Back your claim with evidence.*

Figure 6.1. Cormorant

Figure 6.2. Yellow Perch

ECOLOGY PROBLEM 1

Page 2: More Information

Where's Percho?

The double-crested cormorants were only resting on Loon Lake during their spring migration. However, the birds remain at Les Cheneaux Islands throughout the summer and are very active during the yellow perch spawning period.

In the 1970s, there were 200 cormorants on the Great Lakes, but in 2000, there were approximately 600,000 (Figure 6.3). The increase in recent years seems to be due to the removal of chemical contamination (i.e., DDT) from their environment. The protection provided by a clean environment and by the Migratory Bird Treaty Act has allowed their numbers to increase. When fishermen see such high numbers, they want something done to return the numbers to lower levels because they believe that the birds are eating "their" fish. Is there cause for alarm? What should be done with the cormorant population?

Your Challenge: *Determine if the fishermen of Loon Lake should be concerned with the presence of the fish-eating birds. Back your claim with evidence.*

Figure 6.3. Group of Cormorants

ECOLOGY PROBLEM 1

Page 3: Resources

Where's Percho?

1. *Double-Crested Cormorants in Michigan: A Review of History, Status, and Issues Related to Their Increased Population.* Michigan Department of Natural Resources, Report No. 2, 2005. Available at *www.michigan.gov/documents/Cormorant_Report_136470_7.pdf.*

2. *The Rise of the Double-crested Cormorant on the Great Lakes: Winning the War Against Contaminants*. Environment Canada, Great Lakes Fact Sheet, 1995. Available at *http://publications.gc.ca/collections/Collection/En40-222-2-1995E.pdf.*

3. Michigan Cormorant National Environmental Policy Act. U.S. Fish and Wildlife Service, 2006. Available at *www.fws.gov/Midwest/NEPA/MICormorantNEPA.*

ECOLOGY PROBLEM 1

Teacher Guide

Where's Percho?

Problem Context

The appearance of predatory species, either through human intervention or natural events, sometimes raises concerns by sportsmen about the impact of competition from predators. The increasing number of double-crested cormorants in the Great Lakes region represents such a change. Cormorants have been seen feeding on yellow perch, a popular game fish. Some fishermen have expressed concerns that allowing cormorants to expand their range may reduce the population of perch, thus affecting the quality and quantity of fish available for anglers.

Related Contexts

Similar conflicts between sportsmen and game management staff have occurred in (to cite just two examples) the release of river otters in northern Indiana wetlands and lakes and release of wolves into Yellowstone National Park.

Another source of conflict involves the competition between game species and protected or managed species. This type of case is represented by the dispute between hunters and game management agencies regarding the removal of ring-necked pheasant (an introduced game species) from habitat set aside for the greater prairie chicken (an endangered native species adversely affected by pheasants' nest parasitism) in southern Illinois.

> *Model response:* The increase in double-crested cormorants is actually the return of a species once found in the Great Lakes in large numbers, but reduced in number by hunting and decreased reproduction because of high levels of DDT in the birds. Although cormorants do eat perch, they do not eat only perch. Cormorants will feed on whatever fish species are available at any given time. It is likely that cormorants are partially responsible for a decline in the quality and quantity of fish in the lake, but there are other factors as well, including water quality, overfishing, and changes in the food chain in the Great Lakes resulting from invasive species such as the zebra mussel that reduce the amount of plankton available for the baitfish on which perch feed. Eliminating the cormorant is unlikely to prevent the decline in the perch fishery.

ECOLOGY PROBLEM 1

Assessment

Where's Percho?

Transfer Task

The Bonneville Dam is found on the Columbia River, on the border between Oregon and Washington. A large fish ladder allows Chinook salmon to swim upstream to spawn. The salmon are endangered and are protected as an economically important species. The salmon are also highly valued by Native American tribes in the region.

For the past few years, male California sea lions, a protected marine mammal, have been swimming more than 100 miles up the river to spend the spring in the waters below the dam, gorging on the easy-to-catch salmon in the mouth of the fish ladder. Some groups wish to kill the sea lions to protect the salmon. Other groups oppose the proposal, citing the sea lions' protected status.

(*Important information:* Adult salmon feed on herring in the Pacific Ocean. Young salmon feed on detritus, small invertebrates, and smaller fish as they develop, depending on the size of the salmon. Bears, orcas, and eagles are among the species that feed on salmon. Sea lions feed on salmon but have only recently begun feeding below the dam where the salmon are easy to catch.)

If the sea lions are left in the Bonneville Dam area (Figure 6.4), what effect will the sea lions have on the salmon and the rest of the food web that supports the salmon? Develop a model that will help the interested parties determine if the current situation is stable or if it will result in change. Note that one way that ecologists model ecosystems is with a concept map. This type of map draws all of the living and nonliving parts of the ecosystem in circles and labels arrows between the circles to indicate the relationships among parts.

Figure 6.4. Sea Lions and Bonneville Dam

Model Response: A concept map of the salmon's ecosystem is shown in Figure 6.5 [SEP 2: Developing and Using Models]. The thickest arrow shows the change in the system—sea lions preying on salmon at the dam [CC 7: Stability and Change]. If this greatly reduces the number of adult salmon that go upriver to spawn, every component in the ecosystem connected to salmon will be affected either directly or indirectly. If the sea lions continue swimming up the Columbia River to feed on salmon, the salmon population will decrease. As the sea lions eat adult salmon trying to pass the dam, fewer salmon will reach the spawning areas upstream, and fewer young salmon will hatch to replace the adult population. Over time, this could drastically reduce the number of salmon. This is not likely to cause *all* the salmon to die off though, because as the salmon become scarcer, the sea lions will not eat as many and may quit following the salmon upstream.

Figure 6.5. Concept Map of Salmon Ecosystem

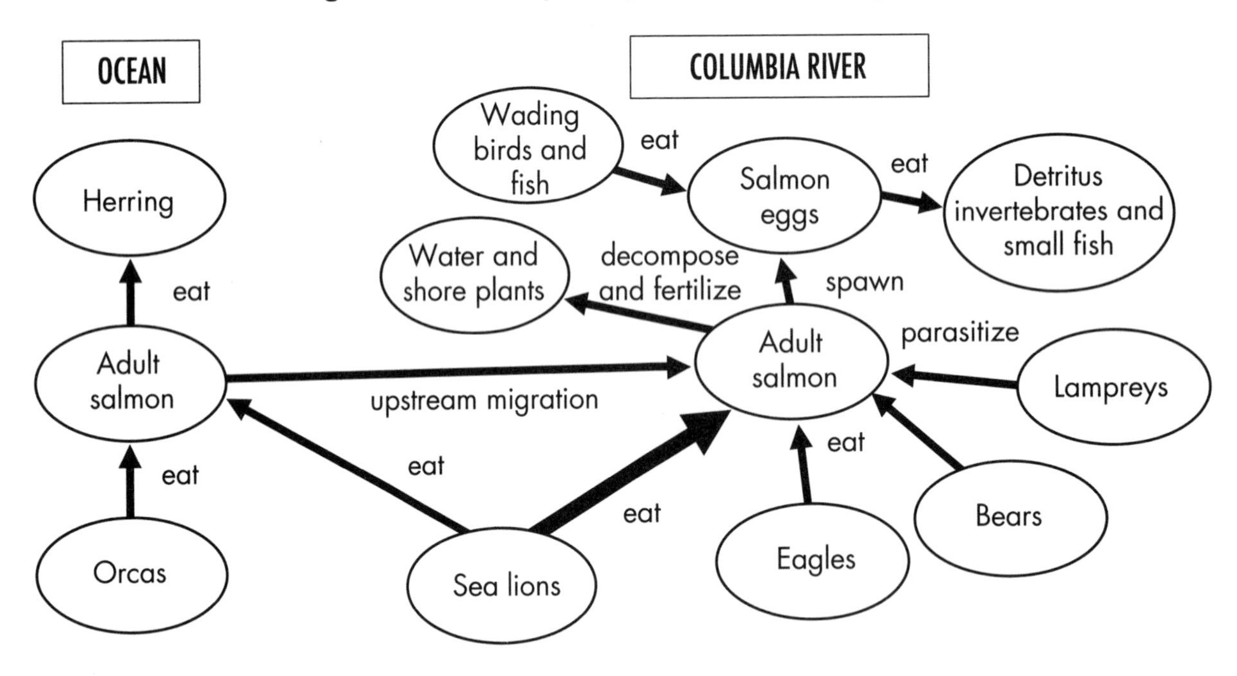

Model response: The decline in salmon will affect other species as well. Animal populations that feed on the salmon farther upstream will decrease. This may include bald eagles, bears, fox, and seagulls. A decline in salmon would also cause a decline in species such as the lamprey, which are parasites feeding on salmon. Wading birds and smaller fish that eat salmon eggs may also decline as that food source becomes scarcer. Plants along and in the river may also decline. The adult fish that die and decay after they spawn provide a rich source of fertilizer as they decompose, and removing this source of nitrogen and phosphate may have an impact on trees, shrubs, grasses, and aquatic plants.

Species that depend on the salmon farther downstream or in the ocean would also be affected, either declining or finding other species to eat. This may include orcas, sharks, and other large predatory fish.

Ecology Problem 2: Lake Michigan—A Fragile Ecosystem

Alignment With the *NGSS*

PERFORMANCE EXPECTATIONS	• *MS-LS2-1:* Analyze and interpret data to provide evidence for the effects of resource availability on organisms and populations of organisms in an ecosystem. • *MS-LS2-2:* Construct an explanation that predicts patterns of interactions among organisms across multiple ecosystems. • *MS-LS2-4:* Construct an argument supported by empirical evidence that changes to physical or biological components of an ecosystem affect populations. • *MS-ESS3-3:* Apply scientific principles to design a method for monitoring and minimizing a human impact on the environment. • *MS-ESS3-4:* Construct an argument supported by evidence for how increases in human population and per-capita consumption of natural resources impact Earth's systems.
SCIENCE AND ENGINEERING PRACTICES	• Asking Questions and Defining Problems • Developing and Using Models • Constructing Explanations and Designing Solutions • Engaging in Argument From Evidence • Obtaining, Evaluating, and Communicating Information
DISCIPLINARY CORE IDEAS	• *LS2.A-MS Interdependent Relationships in Ecosystems:* Predatory interactions may reduce the number of organisms or eliminate whole populations of organisms. Mutually beneficial interactions, in contrast, may become so interdependent that each organism requires the other for survival. Although the species involved in these competitive, predatory, and mutually beneficial interactions vary across ecosystems, the patterns of interactions of organisms with their environments, both living and nonliving, are shared. • *LS2.C-MS Ecosystem Dynamics, Functioning, and Resilience:* Ecosystems are dynamic in nature; their characteristics can vary over time. Disruptions to any physical or biological component of an ecosystem can lead to shifts in all its populations. • *LS4.C-HS Adaptation:* Changes in the physical environment, whether naturally occurring or human induced, have thus contributed to the expansion of some species, the emergence of new distinct species as populations diverge under different conditions, and the decline—and sometimes the extinction—of some species. • *ESS3.C-MS Human Impacts on Earth Systems:* Human activities have significantly altered the biosphere, sometimes damaging or destroying natural habitats and causing the extinction of other species. But changes to Earth's environments can have different impacts (negative and positive) for different living things.
CROSSCUTTING CONCEPTS	• Cause and Effect: Mechanism and Explanation • Systems and System Models

Keywords and Concepts

Competition, food webs, invasive species

Problem Overview

Commercial fishermen have noticed declining numbers and smaller sizes of the fish they catch in Lake Michigan, and they want to know how the government's stocking program is affecting their catch. Full-color versions of all the problem's images are available on the book's Extras page at *www.nsta.org/pbl-lifescience*.

ECOLOGY PROBLEM 2

Page 1: The Story

Lake Michigan—A Fragile Ecosystem

Since the early 1970s, sport fishing for salmon has been excellent along the entire east coast of Lake Michigan. Because of the fishing, a charter boat industry has succeeded and brought economic development to many western Michigan port cities and villages. This $4 billion commercial and sport fishing industry has grown around the Chinook, or king, salmon (*Oncorhynchus tshawytscha;* Figure 6.6).

But for the past two years, the king salmon caught in Lake Michigan have been noticeably reduced in number and size. The Michigan Department of Natural Resources (DNR) has decided to reduce the number of salmon stocked in Lake Michigan. The act concerns those whose livelihood is tied to fishing.

Your Challenge: *Help find out what could be happening to the fish and if people should be concerned. Provide evidence for your claim.*

Figure 6.6. Chinook, or King, Salmon

ECOLOGY PROBLEM 2

Page 2: More Information

Lake Michigan—A Fragile Ecosystem

In 1966, the Michigan DNR stocked coho, or silver, salmon (*Oncorhynchus kisutch*) and Chinook (king) salmon to control the alewife (*Alosa pseudoharengus;* Figure 6.7) population. In the 1960s, alewives made up 90% of the Lake Michigan biomass, but in recent years their numbers have been decreasing. In the early 1990s, Lake Michigan had a biomass of 220 million pounds of alewife. In 1998, that biomass was down to 132 million pounds. In 2005, it was only 55 million pounds. In 1996, alewives made up 43% of the Lake Michigan forage base. Bloater chubs made up 51% and rainbow smelt made up 5%. Because the alewife is the preferred prey of the king salmon, it seems that the number of kings will follow the decline of the prey population.

Five years ago, a 4-year-old king salmon from Lake Michigan would weigh 20 pounds, but now it weighs only 8 pounds. For the past 3 years, 15 million king salmon smolts have been added to Lake Michigan. This year, the stocking rate will be 50% of that. Is the Lake Michigan ecosystem going to continue to change, and what might be causing these changes?

Your Challenge: *Help find out what could be happening to the fish and if people should be concerned. Provide evidence for your claim.*

Figure 6.7. Alewife

ECOLOGY PROBLEM 2

Page 3: Resources

Lake Michigan—A Fragile Ecosystem

1. Local DNR fishery biologists
2. Michigan DNR fish population estimates
3. "The Lake Michigan Fishery: Balancing the Future." 2005. Available at *www.michigan.gov/documents/LakeMichiganBriefingPaperSept05_135869_7.pdf.*
4. "Paradise in Peril," by Dan Egan. *Milwaukee-Wisconsin Journal Sentinel*, December 12, 2004. Available at *www.jsonline.com/news/wisconsin/98940344.html.*
5. "Status of Salmonines in Lake Michigan, 1985–2007." Available at *www.in.gov/dnr/fishwild/files/fw-SWG_2008_Report_data1985to2007_To_the-LMC.pdf.*

ECOLOGY PROBLEM 2

Teacher Guide

Lake Michigan—A Fragile Ecosystem

Problem Context

Over the past few years, the number and size of king salmon caught in Lake Michigan have been declining dramatically. This has had a negative effect on the commercial fishing industry all around Lake Michigan.

The Michigan DNR and the U.S. Fish and Wildlife Service have decided to reduce the number of king salmon stocked in Lake Michigan each year by about 50%. Commercial charter fishermen are concerned that this will decrease the number of fish and their income from fishing even further. To understand how the reduced number of stocked salmon will affect this industry, it is important to understand why the population of salmon has declined and how stocking fewer fish may affect the population in the future.

Related Contexts

Similar management problems have emerged when governments make decisions about dealing with game species that may be overpopulated. In Indiana, the state faced strong opposition to allowing controlled hunting of deer inside several state parks when populations and the health of deer became a problem. Studies of moose populations on Isle Royale in Lake Superior also showed that understanding changes in the food supply of moose was crucial to understanding changes in moose populations.

> *Model response:* When the zebra mussel was accidently introduced into the Great Lakes, the food chain in Lake Michigan changed. The zebra mussel is a very efficient filter feeder. As the number of zebra mussels increased, the water in Lake Michigan became clearer, with less plankton in the lake for the small baitfish to eat. The alewife population, the fish that is the primary food source for salmon and steelhead trout, has declined dramatically. Since the 1950s, the salmon population in Lake Michigan has relied on restocking to maintain population levels. As the alewife population has declined, salmon have less food. Rather than going extinct, the fish are simply not growing as large. This explains the lower quality of fish caught by the sport fishing industry. The state's decision to reduce stocking by 50% is intended to match the salmon population with the new carrying capacity (K) of the lake. There will be fewer fish available, but the fish in the lake will be able to grow larger.

ECOLOGY PROBLEM 2

Assessment

Lake Michigan—A Fragile Ecosystem

Transfer Task

Wildlife managers in northern Indiana have been working to monitor and control populations of bighead carp (*Hypophthalmichthys nobilis*) that have found their way into the Wabash River. These fish are originally from Asia, and their population has begun to increase rapidly. Like other native carp, they feed on decaying material on the bottom of the river, and they tend to stir up a lot of silt in the water.

Describe how the introduction of bighead carp may affect the ecosystem if they continue to increase their numbers.

> *Model Response:* The bighead carp is competing with native carp for food. Because there is a limited carrying capacity for fish in any given niche, the increase in competition means both species will have fewer individuals than if one of the competitors is not present. The rising population of bighead carp suggests that the invasive species is outcompeting the native carp. There are likely to be other species in the Wabash River that either feed on native carp or benefit from the effect of the native carp on the environment. If the bighead carp population continues to increase, those species may also decline.
>
> Because the bighead carp stir up a lot of silt, the cloudy water will also likely make it harder for native algae and aquatic plants to survive because less sunlight will reach the plants. This will eventually disrupt most of the native food webs in the river, and the entire ecosystem could be threatened.

ECOLOGY PROBLEM 3: BOTTOM DWELLERS

Alignment With the *NGSS*

PERFORMANCE EXPECTATIONS	• *MS-LS2-1:* Analyze and interpret data to provide evidence for the effects of resource availability on organisms and populations of organisms in an ecosystem. • *MS-LS2-2:* Construct an explanation that predicts patterns of interactions among organisms across multiple ecosystems. • *MS-LS2-4:* Construct an argument supported by empirical evidence that changes to physical or biological components of an ecosystem affect populations. • *MS-ESS3-3:* Apply scientific principles to design a method for monitoring and minimizing a human impact on the environment. • *MS-ESS3-4:* Construct an argument supported by evidence for how increases in human population and per-capita consumption of natural resources impact Earth's systems.
SCIENCE AND ENGINEERING PRACTICES	• Asking Questions and Defining Problems • Constructing Explanations and Designing Solutions • Engaging in Argument From Evidence • Obtaining, Evaluating, and Communicating Information
DISCIPLINARY CORE IDEAS	• *LS2.A-MS Interdependent Relationships in Ecosystems:* Predatory interactions may reduce the number of organisms or eliminate whole populations of organisms. Mutually beneficial interactions, in contrast, may become so interdependent that each organism requires the other for survival. Although the species involved in these competitive, predatory, and mutually beneficial interactions vary across ecosystems, the patterns of interactions of organisms with their environments, both living and nonliving, are shared. • *LS2.C-MS Ecosystem Dynamics, Functioning, and Resilience:* Ecosystems are dynamic in nature; their characteristics can vary over time. Disruptions to any physical or biological component of an ecosystem can lead to shifts in all its populations. • *LS4.C-HS Adaptation:* Changes in the physical environment, whether naturally occurring or human induced, have thus contributed to the expansion of some species, the emergence of new distinct species as populations diverge under different conditions, and the decline—and sometimes the extinction—of some species. • *ESS3.C-MS Human Impacts on Earth Systems:* Human activities have significantly altered the biosphere, sometimes damaging or destroying natural habitats and causing the extinction of other species. But changes to Earth's environments can have different impacts (negative and positive) for different living things.
CROSSCUTTING CONCEPTS	• Cause and Effect: Mechanism And Explanation • Systems and System Models

Keywords and Concepts

Eutrophication, water quality indicators, water pollution, human impacts on the environment

Problem Overview

Residents on a small lake are concerned about the loss of a once-common insect from the ecosystem. What might be the cause? Full-color versions of all the problem's images are available on the book's Extras page at *www.nsta.org/pbl-lifescienc*.

ECOLOGY PROBLEM 3

Page 1: The Story

Bottom Dwellers

Loon Lake, a 63-acre lake located in Montcalm County, Michigan, has not experienced a mayfly (*Hexagenia limbata*) emergence in the past four years. Populations of the mayfly seem to have crashed at Loon Lake, but neighboring Crystal Lake, a 724-acre lake, has had large emergences. Both lakes are all-sports lakes with many dwellings along their shorelines. Crystal Lake has dwellings completely surrounding the water, whereas Loon Lake has cottages only on three sides. The south shore has not been developed because it is a wetland. Figure 6.8 shows a mayfly nymph, and Figure 6.9 shows an adult mayfly.

Your Challenge: *Help the homeowners discover what is happening to the Loon Lake ecosystem that would cause a decline in the mayfly population.*

Figure 6.8. Mayfly Nymph

Figure 6.9. Adult Mayfly

ECOLOGY PROBLEM 3

Page 2: More Information

Bottom Dwellers

Both lakes are "typical" southern Michigan eutrophic lakes. Loon Lake receives groundwater from Fish Creek watershed, and its water drains into Crystal Lake. The houses on Loon Lake are smaller and older, and they have septic tanks. The homes on Crystal Lake were built more recently and are connected to a municipal sewer system.

The bottom of both lakes consists mainly of marl and calcium carbonate, which provides a good substrate for burrowing insects such as *H. limbata*. Loon Lake has large emergences of the bloodworm (or "buzzer") midge (*Chironomus plumosus*) in the spring and fall. This midge also burrows in the bottom sediment. It is not as abundant in Crystal Lake as it is in Loon Lake.

Both lakes have healthy and large populations of bluegills, largemouth bass, and northern pike. Loon Lake also has a few very large yellow perch and black crappies, but small individuals of each species are not present. Carp (Figure 6.10) and other minnows are also abundant in both lakes. Tree swallows (Figure 6.11) are often seen feeding on emerging insects.

What could cause the difference in the insect populations on Loon and Crystal Lakes?

Your Challenge: *Help the homeowners discover what is happening to the Loon Lake ecosystem that would cause a decline in the mayfly population.*

Figure 6.10. Carp

Figure 6.11. Tree Swallow

ECOLOGY PROBLEM 3

Page 3: Resources and Investigations

Bottom Dwellers

1. "Benthic Macroinvertebrates," in *Hoosier Riverwatch Volunteer Stream Monitoring Training Manual 2015*. Indiana Department of Environmental Management, 2015. Available at *www.in.gov/idem/riverwatch/files/volunteer_monitoring_manual.pdf.*

2. "Blood Midges." Available at *www.flyline.com/entomology/blood_midges.*

Lake or Stream Water Sampling Investigation

MATERIALS

- Water sample–collecting jars
- D-frame or kick seine nets
- Dissolved oxygen (DO) test kits (e.g., Hach or Ward's Science Snap Test kits, probeware)
- Thermometer or digital temperature probe
- DO and invertebrate sample water quality (Q-value) charts (available from volunteer water quality manuals in most states)
- Plastic container for examining macroinvertebrates
- Benthic macroinvertebrate guide or dichotomous key (available from volunteer water quality manuals in most states)

SAFETY PRECAUTIONS

- Use a chemical disposal container to dispose of used DO test samples.
- Keep a flotation device (e.g., boat cushion) handy while students are collecting samples.
- Return animals collected to the stream or lake where they were found.

PROCEDURE

1. Students can collect water samples from a local stream or lake to test for DO. If possible, compare two different bodies of water or the same source at different times of the year.

2. Measure and record the temperature of the water in degrees Celsius.

3. Follow the instructions in the selected test kit to find and record the total DO (in mg/L).

4. Use the water quality index charts and the total DO and temperature data to find the percent saturation (% saturation) for your samples.

5. Use the Q-value chart to find the water quality index for the sample.

6. Use the D-frame or kick seine nets to collect samples of the insect larvae and other invertebrates in the lake or stream.

7. Place the samples in plastic containers with some water. Use the benthic macroinvertebrate guide or dichotomous key to identify species collected. Record data in the invertebrate sample water quality chart.

8. Use the invertebrate sample water quality index to calculate the quality value for the invertebrate sample.

9. Analyze the data to identify patterns in DO values and diversity of invertebrates in various samples.

ECOLOGY PROBLEM 3

Teacher Guide

Bottom Dwellers

Problem Context

Loon Lake and Crystal Lake are located very near each other. Crystal Lake has a normal population of mayflies, but there have been no mayflies in Loon Lake for four years. Both lakes have homes along the shoreline, but homes on Crystal Lake are connected to a sewer system rather than having septic systems. The lakes are very similar, but the oxygen levels in Loon Lake are no longer high enough for mayflies to thrive. One of the key factors affecting oxygen levels is that the homes on Loon Lake have septic systems that have been in place for decades. Organic chemicals leeching from the septic tanks add nutrients to Loon Lake, which alters the quality of water.

Related Context

Many species of invertebrates are indicators of water quality. Mayflies, stoneflies, caddisflies, dobsonflies, riffle beetles, water pennies, and right-handed snails all need high dissolved oxygen (DO) levels to survive. Other species, such as bloodworm midges and left-handed snails, can survive in lakes with little DO. Some kinds of fish can also indicate DO levels (trout) or low DO levels (perch, carp, shad). These populations can be used to identify lakes that may have natural or human-made factors that reduce the DO levels in the water. Factors that reduce DO include higher temperatures, less plant growth, and increased oxygen consumption by high levels of bacteria decomposing organic materials.

> *Model response:* On Crystal Lake, the sewer system takes wastes from the homes away from the lake and prevents organic wastes from entering the watershed. The homes on Loon Lake, with their older septic systems, leech organic wastes into the groundwater, and the wastes eventually find their way into the water. The organic wastes provide a food source for bacteria in the lake. As the bacteria break down the wastes, they consume a great deal of oxygen. Water tests in the lakes would show that the percent saturation of DO in Loon Lake is lower than in Crystal Lake.
>
> Mayflies are a species of insect larva that can indicate water quality. They are very intolerant of pollution and low DO levels. Water quality indexes list the mayfly as a Group 1 ("Intolerant") species. As DO levels drop, the mayflies are unable to survive.

Other species like the gray midge and bloodworm midge are Group 3 ("Fairly Tolerant") and Group 4 ("Very Tolerant"), respectively, meaning they can tolerate much lower DO levels and higher levels of pollution. This explains why the mayfly is not present and the midges are still found in Loon Lake. Over time, the change in diversity of insect larvae can be expected to have an impact on the population of some fish, and species like perch and pike that need high oxygen levels will likely disappear from Loon Lake.

ECOLOGY PROBLEM 3

Assessment

Bottom Dwellers

Transfer Task

Figure 6.12. Fir Trees

Isle Royale, the largest island in Lake Superior, provides biologists with a unique system for studying the interactions between different trophic levels. Back in 1959, Dr. Durward Allen of Purdue University began a study of the simple food chain consisting of producers, a single large herbivore, and a single large predator, the gray wolf (*Canis lupus*).

The island had a large stand of balsam fir (Figure 6.12) until the early 1900s, when moose (Figure 6.13) colonized the park by swimming to the island. After the establishment of the moose, the balsam fir declined from 46% of the overstory to about 5% today. Nearby islands that are inaccessible to moose continue to have large fir stands, so some scientists have suggested that feeding activities of moose have caused the decline of the fir on Isle Royale. Through the years, significant fluctuations have been observed in the densities of the wolf and moose populations and the growth rates of balsam fir.

Figure 6.13. Moose

Figure 6.14 (p. 130) presents data for several key variables related to the populations of wolves, moose, and fir trees. The actual evapotranspiration rate (AET) serves as an index of the amount of water available for plant growth and is strongly tied to primary productivity.

Scientists are wondering which factor has the biggest influence on the population of moose and wolves on Isle Royale. Using what you know about food chains and food webs, draw a model of this system. Suggest at least one explanation for the fluctuations in these populations. Be as detailed as possible.

Model Response: Figure 6.15 shows a model of the Isle Royale ecosystem [SEP 2: Developing and Using Models]. If large enough, change in any part of the system will affect the other parts of the system that are directly related to it. Eventually, changes in these related parts may spread to more parts [CC 7: Stability and Change].

Figure 6.14. Wolf, Moose, and Fir Populations

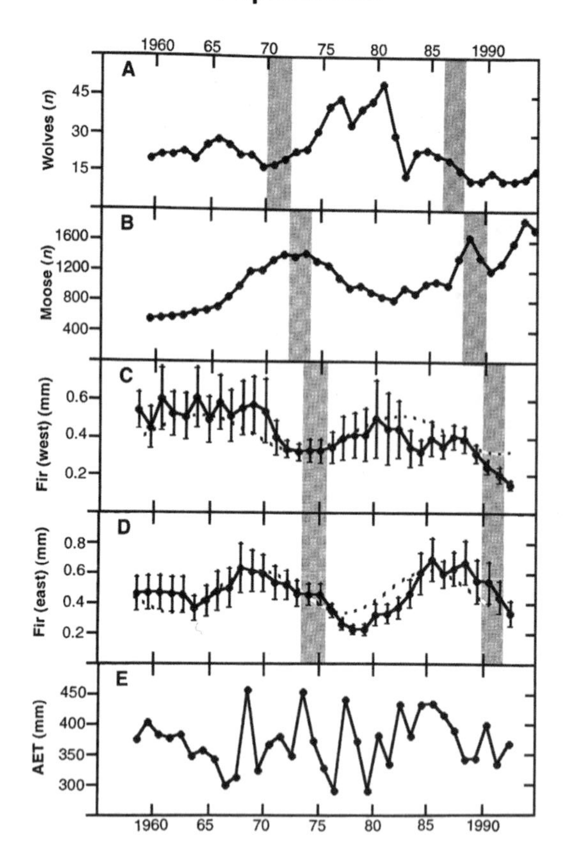

Source: McLaren and Peterson 1994.

Populations will fluctuate around the carrying capacity (K). The carrying capacity of any ecosystem depends on the amount of space or shelter, water, and food available for the species being studied. On Isle Royale, because differences in rainfall, snowfall, and temperature affect the growth of the trees, the population will see corresponding variations. After drier years, fewer trees will exist. The fir population then affects the moose because the moose eat fir trees. If fir populations decline, over a period of years the moose will decline. If moose populations are high, they also eat many trees, reducing the fir population, which leads to the same decline in moose. The change in the moose population lags—the decline in moose takes years to appear after the decline in fir trees because the moose have a relatively long life and it takes them a long time to generate new young. It also takes time before the decline in fir trees leads to starvation among the moose.

A similar pattern shows up in the populations of moose and wolves. A rise in the moose population will lead to a rise in the wolf population, but there is a lag of a couple of years. Likewise, when moose populations decline, the wolves will decline a couple of years later. The data in the study (McLaren and Peterson 1994) show these patterns occurring repeatedly over many years [SEP 4: Analyzing and Interpreting Data].

Figure 6.15. Model of Isle Royale Ecosystem

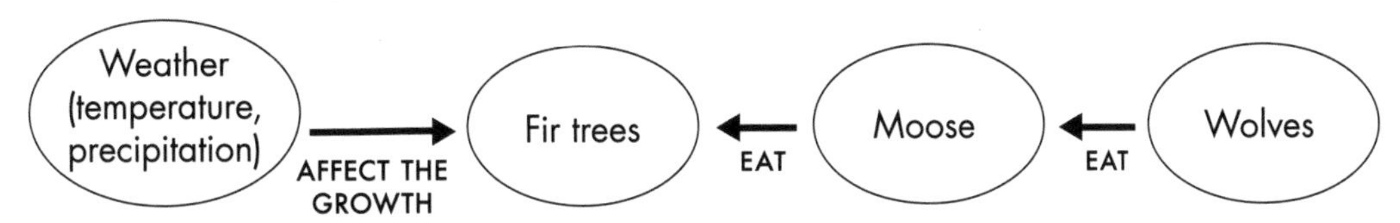

Reference

McLaren, B. E., and R. O. Peterson. 1994. Wolves, moose and tree rings on Isle Royale. *Science* 266 (5190): 1555–1558.

Ecology Problem 4: Bogged Down

Alignment With the *NGSS*

PERFORMANCE EXPECTATIONS	• *MS-LS2-1:* Analyze and interpret data to provide evidence for the effects of resource availability on organisms and populations of organisms in an ecosystem. • *MS-LS2-2:* Construct an explanation that predicts patterns of interactions among organisms across multiple ecosystems. • *MS-LS2-4:* Construct an argument supported by empirical evidence that changes to physical or biological components of an ecosystem affect populations.
SCIENCE AND ENGINEERING PRACTICES	• Asking Questions and Defining Problems • Developing and Using Models • Constructing Explanations and Designing Solutions • Engaging in Argument From Evidence • Obtaining, Evaluating, and Communicating Information
DISCIPLINARY CORE IDEAS	• *LS2.C-MS Ecosystem Dynamics, Functioning, and Resilience:* Ecosystems are dynamic in nature; their characteristics can vary over time. Disruptions to any physical or biological component of an ecosystem can lead to shifts in all its populations. • *LS4.C-HS Adaptation:* Changes in the physical environment, whether naturally occurring or human induced, have thus contributed to the expansion of some species, the emergence of new distinct species as populations diverge under different conditions, and the decline—and sometimes the extinction—of some species.
CROSSCUTTING CONCEPTS	• Patterns • Cause and Effect: Mechanism and Explanation • Systems and System Models • Stability and Change

Keywords and Concepts

Ecological succession, plant dispersal, competition in plants

Problem Overview

The plants found in a bog are changing over time. What is causing this change? Is there a human-related problem that needs to be addressed? Full-color versions of all the problem's images are available on the book's Extras page at *www.nsta.org/pbl-lifescience*.

ECOLOGY PROBLEM 4

Page 1: The Story

Bogged Down

For the last 20 years, students have been visiting a bog located on the west side of the Rose Lake marsh in Clinton County, Michigan (Figure 6.16). Every year, more and more red maple trees (*Acer rubrum*) have been observed growing inside the stand of tamaracks that border a small pond. It looks as though the red maples will eventually shade the tamaracks and the other bog plants, thereby eliminating them.

Your Challenge: *Help find factors that have caused the changes at the Rose Lake marsh, and predict what we can expect to happen to the bog ecosystem over time.*

Figure 6.16. Rose Lake Bog

Note: A full-color version of this figure is available on the book's Extras page at *www.nsta.org/pbl-lifescience.*

ECOLOGY PROBLEM 4

Page 2: More Information

Bogged Down

A bog is a standing body of water with no underground spring of freshwater to feed it. The water is generally cold, extremely acidic, and low in oxygen. Sphagnum moss grows in the bog and forms a thick mat of floating plants (Figure 6.17). These plants, over time, can fill in the pond or small lake with peat that will eventually be firm enough to support trees.

In the middle of the bog is an area of open water. Around that is a border of mint and cattails, and just a few feet closer to the center is a mat of sphagnum moss and other plants that is so thick in spots a person can walk on top of the mat without falling into the water. The whole mat moves up and down. As a result, these ecosystems are sometimes called *quaking bogs*. This peat bog stage is followed over time by shrubs and tamaracks.

Bogs have very little decomposition of organic matter because sphagnum gives off hydrogen ions, creating a very acidic soil. In this nutrient-poor soil, some plants have adapted by becoming carnivorous. Examples include sundews and pitcher plants (Figure 6.18). These plants trap insects to supplement their photosynthetic diets.

Your Challenge: *Help find factors that have caused the changes at the Rose Lake marsh, and predict what we can expect to happen to the bog ecosystem over time.*

Figure 6.17. Bog Plants

Figure 6.18. Pitcher Plant Flower

Note: Full-color versions of the figures above are available on the book's Extras page at *www.nsta.org/pbl-lifescience*.

ECOLOGY PROBLEM 4

Page 3: Resources

Bogged Down

1. Field trip to Rose Lake (or other bogs)

2. "Plant Succession." 2015. Available at *http://users.rcn.com/jkimball.ma.ultranet/BiologyPages/S/Succession.html.*

3. "Peatlands, Bogs and Fens." Minnesota Department of Natural Resources, 2016. Available at *www.dnr.state.mn.us/snas/coniferous_peatlands.html.*

4. "Volo Bog State Natural Area." Illinois Department of Natural Resources, 2016. Available at *www.dnr.illinois.gov/Parks/Pages/VoloBog.aspx.*

ECOLOGY PROBLEM 4

Teacher Guide

Bogged Down

Problem Context

Rose Lake has a small bog in one section of the wetlands. This bog contains sphagnum moss, sundews, pitcher plants, cranberry and blueberry bushes, and tamarack trees. Red maple trees are beginning to grow in the bog and seem to be indicating a change in the environment.

Related Contexts

Other parts of the Rose Lake property show signs of ecological succession. Areas that were once grassland now have Russian olive, honeysuckle, multiflora rose, dogwood, and other shrubs in them. The lake has become smaller, whereas the cattail marsh has expanded into areas that were once open water. Areas once dominated by shrubs and cattails now have cottonwood and maple trees growing in them. Similar changes can be found near schools in most areas. Examples might include a park or former farm fields that have been left fallow and are changing in predictable stages. The dunes of eastern Lake Michigan are also examples of succession that students may be familiar with. Mount St. Helens and other volcanoes are areas that are being studied to track the changes in ecosystems as ecological succession progresses.

> *Model response:* The Rose Lake bog is a naturally occurring ecosystem that forms when a glacial lake develops in a location in which sphagnum moss grows, creating an acidic environment that influences the types of plants that can survive. The moss grows a mat that covers the water below and provides a base on which other plants can grow. The high acidity inhibits decomposition, so the environment is low in nitrates. The plants that can survive in this habitat are the characteristic species in the bog—sundew, pitcher plant, cranberry, blueberry, and tamarack.
>
> As the bog ages, more sediments build up and the ecosystem undergoes the natural change we call ecological succession. In a bog, as the lake below the moss fills in, the edges of the bog become less acidic, and seeds from other types of plants can survive. Cattails, maple trees, white pine trees, and other plants begin to outcompete the shorter bog species for sunlight and other resources. Eventually, the bog will become either a marsh or a mixed forest. These changes are normal and occur without human interference. The changes seen at Rose Lake are simply evidence of ecological succession.

ECOLOGY PROBLEM 4

Assessment

Bogged Down

Transfer Task

Andrew and Jeff like to play next to the lake just a few hundred yards from their house. The lake is not very big, but since they were much younger, the boys would wade at the edge of the lake for frogs and turtles, watch dragonflies near the cattails by the shore, and fish from a small pier along one side of the lake. Their dad even took them out in a rowboat to fish in the middle of the lake.

The water used to be about 14 feet deep, but now, about 8 years later, the deepest part of the lake is only about 12 feet deep. The area where they used to wade is mostly shrubs and sedges, and the cattails reach almost halfway to the middle of the lake. There is definitely less open water in the middle.

What could possibly be shrinking the lake the boys enjoyed so much? Is there something wrong with the environment that might cause this change?

> *Model Response:* The lake is shrinking (or, more accurately, filling in) over time because of a natural process called ecological succession. The plants and animals in any lake will eventually die and decompose, adding nutrients and other materials to the mud and silt in the lake. Soil from the surrounding land also ends up in the lake because of erosion from wind and water. As the lake ages, the water will get shallower, and plants normally found only on the margins of the lake will spread toward the center. Given enough time, the lake will become a marsh or swamp, and eventually a forest.
>
> The changes the boys are noticing are not a sign of an environmental problem. They are simply the natural sequence of changes that take place in most lakes.

Ecology Problem 5: The Purple Menace

Alignment With the *NGSS*

PERFORMANCE EXPECTATIONS	• *MS-LS2-1:* Analyze and interpret data to provide evidence for the effects of resource availability on organisms and populations of organisms in an ecosystem. • *MS-LS2-2:* Construct an explanation that predicts patterns of interactions among organisms across multiple ecosystems. • *MS-LS2-4:* Construct an argument supported by empirical evidence that changes to physical or biological components of an ecosystem affect populations. • *MS-ESS3-3:* Apply scientific principles to design a method for monitoring and minimizing a human impact on the environment. • *MS-ESS3-4:* Construct an argument supported by evidence for how increases in human population and per-capita consumption of natural resources impact Earth's systems.
SCIENCE AND ENGINEERING PRACTICES	• Asking Questions and Defining Problems • Constructing Explanations and Designing Solutions • Engaging in Argument From Evidence • Obtaining, Evaluating, and Communicating Information
DISCIPLINARY CORE IDEAS	• *LS2.A-MS Interdependent Relationships in Ecosystems:* Predatory interactions may reduce the number of organisms or eliminate whole populations of organisms. Mutually beneficial interactions, in contrast, may become so interdependent that each organism requires the other for survival. Although the species involved in these competitive, predatory, and mutually beneficial interactions vary across ecosystems, the patterns of interactions of organisms with their environments, both living and nonliving, are shared. • *LS2.C-MS Ecosystem Dynamics, Functioning, and Resilience:* Ecosystems are dynamic in nature; their characteristics can vary over time. Disruptions to any physical or biological component of an ecosystem can lead to shifts in all its populations. • *LS4.C-HS Adaptation:* Changes in the physical environment, whether naturally occurring or human induced, have thus contributed to the expansion of some species, the emergence of new distinct species as populations diverge under different conditions, and the decline—and sometimes the extinction—of some species. • *ESS3.C-MS Human Impacts on Earth Systems:* Human activities have significantly altered the biosphere, sometimes damaging or destroying natural habitats and causing the extinction of other species. But changes to Earth's environments can have different impacts (negative and positive) for different living things.
CROSSCUTTING CONCEPTS	• Cause and Effect: Mechanism and Explanation • Systems and System Models

Keywords and Concepts

Competition, invasive species

Problem Overview

Purple loosestrife is spreading in a cattail marsh. What effect will this have on the ecosystem? Can it be controlled? Full-color versions of all the problem's images are available on the book's Extras page at *www.nsta.org/pbl-lifescience*.

ECOLOGY PROBLEM 5

Page 1: The Story

The Purple Menace

During a visit to Rose Lake in 1996, a single plant of purple loosestrife (*Lythrum salicaria;* Figure 6.19) was seen at the boat access. In 2006, 50 plants could be seen from the same location. The plants seem to have scattered throughout the marsh and seem to be very successful as neighbors to the dominant cattails.

Your Challenge: *Predict what effect purple loosestrife will have on this wetland ecosystem in the future. What, if anything, should be done to control the purple loosestrife?*

Figure 6.19. Purple Loosestrife

Note: A full-color version of this figure is available on the book's Extras page at *www.nsta.org/pbl-lifescience.*

ECOLOGY PROBLEM 5

Page 2: More Information

The Purple Menace

Rose Lake, located in Clinton County, Michigan, shares a boundary with Rose Lake Wildlife Research Area. It is a public access lake that is managed by the Michigan Department of Natural Resources. It is a very old lake, having been formed as the last glacier retreated from southern Michigan. The lake basin has gradually filled, and today there is a small open-water area surrounded by an extensive stand of cattails that supports a very diverse wetland community. The common cattail (*Typha latifolia;* Figure 6.20) is a keystone species that supports many animal species. Its rhizomes also allow the cattail to grow out into the open-water area, which over time will completely eliminate the aquatic ecosystem.

The niche of purple loosestrife in North American wetlands is very similar to its niche in its native lands in Eurasia. Since it was introduced to North America in the 19th century, it has caused problems in other wetland communities. What might happen to the Rose Lake marsh, and what could be done to reduce the spread of this plant?

Your Challenge: *Predict what effect purple loosestrife will have on this wetland ecosystem in the future. What, if anything, should be done to control the purple loosestrife?*

Figure 6.20. Cattails

ECOLOGY PROBLEM 5

Page 3: Resources

The Purple Menace

1. "Purple Loosestrife in the Great Lakes Region." Great Lakes Information Network, 2016. Available at *www.great-lakes.net/envt/flora-fauna/invasive/loosestf.html.*

2. "Purple Loosestrife" (see the sections on control), by Bernd Blossey. Cornell University Ecology and Management of Invasive Plants Program, 2002. *Available at www.invasiveplants.net/plants/purpleloosestrife.htm.*

3. "Purple Loosestrife." Indiana Department of Natural Resources. Available at *www.in.gov/dnr/entomolo/4529.htm.*

4. "Purple Loosestrife in Michigan: Biology, Ecology, and Management." Michigan Sea Grant College Program. Available at *www.miseagrant.umich.edu/downloads/ais/fs-97-501_purple_loosestrife.pdf.*

5. "Beetle Bash Purple Plant Invaders," by Laura Walikainen. MichiganTech News, 2003. Available at *www.admin.mtu.edu/urel/news/media_relations/187.*

ECOLOGY PROBLEM 5

Teacher Guide

The Purple Menace

Problem Context

Purple loosestrife is an invasive species of plant that is spreading through North American wetlands. Stories about the spread of the plant periodically appear in the popular media, so this species may be familiar to participants. In a local wetland at Rose Lake, purple loosestrife has been found and is spreading through the wetland.

Related Contexts

Similar contextual stories involve other invasive species commonly found in the Midwest. These could include garlic mustard, bush honeysuckle, starlings, zebra mussels, Asian grass carp, and ring-necked pheasant. All of these species were introduced to the United States and have quickly adapted and spread through a variety of ecosystems, competing with and sometimes displacing native species.

Another contextualized story would be the introduction of non-native species during the colonization of the Americas. The May 2007 issue of *National Geographic* includes an in-depth description of several invasive species that have had important influences on the ecosystem.

> *Model response:* Purple loosestrife is an invasive species brought to America from Eurasia. It has no native predators that control population growth, and loosestrife is very well adapted to thrive in the same types of wetlands as the native cattail. Because of the root structure and the lack of predators, loosestrife has been able to outcompete the cattails. Once they appear in a wetland, the loosestrife will eventually crowd out most of the native species. Because cattails are a keystone species (many other organisms depend on cattails for food, shelter, etc.), the other species also can no longer survive when cattails disappear. Loosestrife can disrupt entire wetland ecosystems.
>
> There are three main strategies for controlling loosestrife: physical methods, chemical methods, and biological methods. Physical methods include cutting or uprooting the plants. This can work, but it is very labor intensive, and any part of the plant that is left behind or any seeds dropped in the process will grow back. Chemical methods such as herbicides work, but they affect other plants and animals adversely, so they are not recommended. The biological methods include the use of three Eurasian insects

(two types of weevils and one beetle). These species each specialize in eating *only* the leaves, stems, or roots of the purple loosestrife. After extensive testing to ensure they will not eat other native species, scientists have begun using these insects to control loosestrife in Wisconsin, Michigan, and Minnesota. Results have been very promising! The insects do not completely destroy all the loosestrife plants, but they are effective enough in controlling the population that cattails can still survive in the wetlands. This preserves the native ecosystem. In this case, using insects as a biological control seem to be a good idea.

Activity Guide

Nearby wetlands, such as Lake Lansing, that have a large loosestrife population could provide an opportunity for a transect in which students could count the number of species in sample areas to compare diversity with an area like Rose Lake, where loosestrife is only starting to appear.

ECOLOGY PROBLEM 5

Assessment

The Purple Menace

Transfer Task

At Denny's house on the banks of the Looking Glass River in DeWitt, Michigan, there is a strip of natural hardwood vegetation between a narrow lawn and the river. When spring arrived, Denny expected to see a woodland floor full of spring wildflowers but was disappointed when he only found nine stems of trillium in May and later a few wild geraniums. Bush honeysuckle, a flowering shrub native to Russia, is very common in the strip of woods.

Use a model to explain the changes in the diversity of spring flowers that are usually present in mid-Michigan woodlands in early springtime.

Figure 6.21. Woodland Ecosystem Model

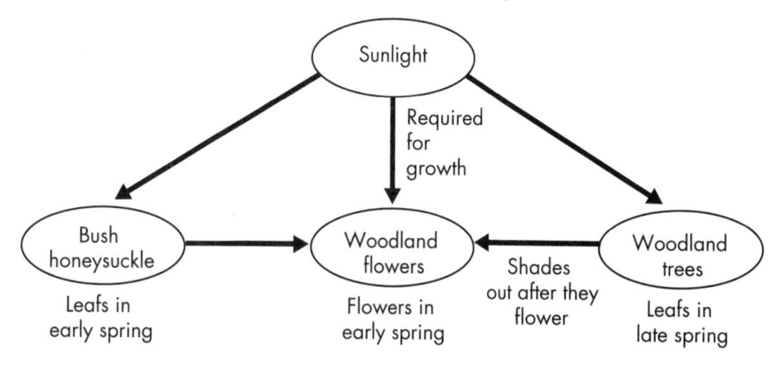

Model Response:

Bush honey-suckle is an invasive species. It is a large shrub that grows to more than 7 feet tall and produces a dense covering of leaves. The plants are perennials and produce leaves and flowers in early spring and small red berries by mid-June.

The native wildflowers found in the woods along Michigan streams are generally only about a foot tall and emerge early in the spring. They blossom in the early spring before leaves open on the taller trees because they need to reproduce before they are shaded by the trees.

When an invasive species such as bush honeysuckle is introduced into a habitat, it often resists local diseases and lacks predators that control its population. This gives it a competitive advantage over local plants. The bush honeysuckle outcompetes the shorter wildflowers. This competitive exclusion allows the honeysuckle to spread and reduces the diversity of plant species in the ecosystem [CC 7: Stability and Change].

Figure 6.21 presents a model showing how honeysuckle interacts with the native woodland flowers [SEP 2: Developing and Using Models].

Ecology Problems: General Assessment

General Question

Scientists have noted that the populations of some species have been declining. In some locations, the diversity of species has decreased, reflecting a complete loss of some species. Explain how various factors—both natural and human-made—may be contributing to the decline of species and the loss of biodiversity.

Model response: The population of a species in an ecosystem is influenced by the availability of limited resources, such as food, space, water, and sunlight. (Examples: number of salmon in Lake Michigan, number of cattails or loosestrife in a marsh, and number of wildflowers and shrubs at the edge of the woods)

Food webs are complex models of food-based interactions between organisms in an ecosystem. A change in one species can influence the population of other species in the web by changing the availability of resources. (Example: removing salmon from the Columbia River and the effect on all the species that eat salmon, including gulls, black bears, and herons)

Introduction of invasive species can result in competitive exclusion of some native species, reducing biodiversity. (Examples: zebra mussels in Great Lakes outcompeting native mussels and purple loosestrife outcompeting cattails)

If certain key species in an ecosystem (keystone species) are eliminated, a wide range of species that depends on the keystone species will be negatively affected. (Example: cattails in a marsh that create habitats for insect larvae, muskrats, amphibians, small fish, and other species that may be eliminated by competition from loosestrife)

Natural changes in the ecosystem (ecological succession) can reduce biodiversity. As environmental conditions change, plants and animals no longer adapted for that habitat do not survive. (Examples: natural changes in a peat bog change soil acidity, allowing for trees to grow that will eventually eliminate bog plants, and loss of diversity in mixed shrublands as they develop into a beech-maple climax forest)

Human-generated pollution and habitat destruction can affect environmental quality, limiting the diversity of species. (Examples: fertilizer runoff leading to algal bloom and eventual loss of oxygen-dependent fish and insects, DDT reducing the number of fish-eating birds of prey, and destruction of forests and wetland for human development)

When more than one species need the same limited resource (light, water, food, space), the species compete. A species with a competitive advantage will increase in population, while other less competitive species will be limited.

Application Questions

APPLICATION QUESTION 1

Shipshewana Lake in Lagrange County, Indiana, is surrounded by farmland, most of it used for livestock. The south shore of the lake is lined with rural homes, each with a septic tank. Organic waste from the homes and farms has entered the lakes through runoff and groundwater. As a result, the lake is no longer as productive for fishing as it once was.

Explain why the lake has changed, including chemical changes and the effect of the environmental changes on the types of organisms that might be found in the lake.

Model response: The organic wastes from homes and farms act as a fertilizer. The septic tanks and animal wastes from the homes and farms are carried into the lake through runoff and cause an imbalance in the nitrogen in the lake, causing rapid growth of algae. When the algae begin to die off, the bacteria that decompose them consume most of the oxygen dissolved in the lake. As a result, fish that need high levels of oxygen, including most game fish, die off [CC 7: Stability and Change].

APPLICATION QUESTION 2

Purple loosestrife is a plant introduced to the United States that thrives in marshes and along the shores of lakes and streams. It has overrun wetlands in the Midwest, eliminating cattails, which are native plants that have been almost completely eliminated in some marshes.

Explain how loosestrife can overrun a marsh and why the cattail population declines. Explain why the change in the plants in the marsh might affect other species in the wetland.

Model response: Purple loosestrife is an invasive species. It lives at the edge of wetlands, a niche very similar to the native cattail plants found in the Midwest. But loosestrife has no natural predators in the United States, so it can grow unchecked. The loosestrife has spread very fast through wetlands, crowding out the cattail plants over time.

This crowding out also affects other species. Cattails are important as a food source for many species of insects, small mammals, and birds. In addition, cattails provide a home for small fish whose babies hide in the root mass of the cattail, for birds such as the marsh wren that depend on the plant for nest sites, and for many other organisms in the wetland ecosystem. The cattail is a keystone species, meaning the marsh ecosystem depends on it. When the cattails are crowded out by the loosestrife, the species that depend on the cattail can no longer survive. This means the loosestrife can disrupt a large part of the biodiversity normally found in these marshes.

APPLICATION QUESTION 3

Figure 6.22 shows two species in an ecosystem, with species A having been there first. The dotted line shows a point in time when species B was introduced into the ecosystem and began to compete with species A for some resource.

Extend the graph showing what will happen to the population of *both* species, assuming both species survive. Then explain why the population sizes will change in the pattern you drew.

Figure 6.22. Graph of Two Populations in Competition

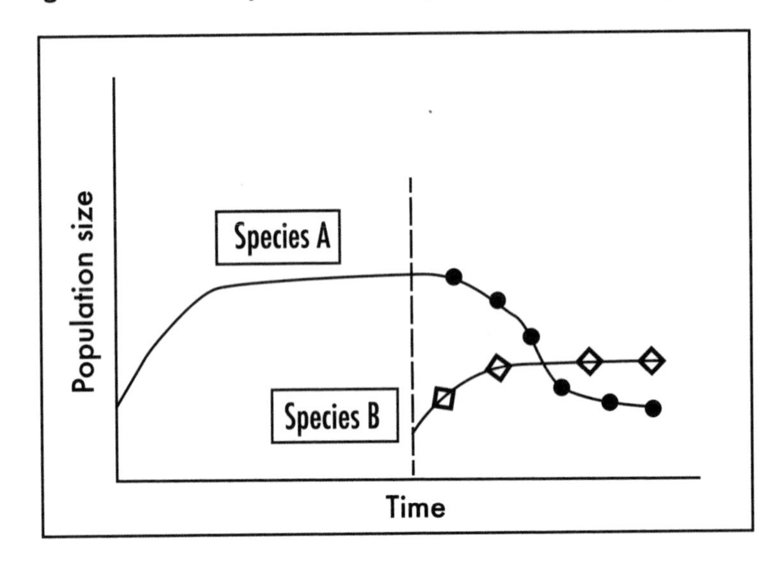

Model response: When species B is introduced, it will increase over time. Because it competes with species A for a limited resource, species A will decline as the resource is used by species B. This trend will continue until both species level off at a new equilibrium. The population of species A may be larger or smaller than species B, depending on which species is better at competing. It is possible one species may be so much better that the other goes extinct. But both species' populations will be smaller than if the other species was not present [SEP 5: Using Mathematics and Computational Thinking; SEP 6: Constructing Explanations and Designing Solutions].

Common Beliefs

Indicate whether the following statements are true (T) or false (F), and explain why you think so (*model responses shown in italics*).

1. When a new species enters an ecosystem, humans must have introduced it. *(F) New species can be introduced to an ecosystem naturally when species spread to new areas. Natural changes in the ecosystem (ecological succession) can reduce biodiversity.*

2. Invasive species affect only the species they compete with for food. *(F) The impact of invasive species can affect multiple species, including competitors, prey, predators, and the species all those affected interact with. The competition is also for space, sunlight, water, shelter materials, and other resources. The introduction of invasive species can result in competitive exclusion.*

3. Plants compete for resources in ways very similar to competition between animals. *(T) Plant competition is similar to competition in animals, though the competition for food focuses on sunlight (for photosynthesis). The structures that adapt to compete differ. The population of a species in an ecosystem is influenced by the availability of limited resources.*

4. A population's size cannot exceed the carrying capacity of its ecosystem. *(F) A population can rise above the carrying capacity for a short time. When the population exceeds the carrying capacity, the population will decline to levels below that number, then climb again. The population of a species in an ecosystem is influenced by the availability of limited resources. When more than one species need the same limited resource (light, water, food, space, etc.), the species compete.*

5. Growth of human populations changes over time in patterns very similar to those of other animal species. *(T) Human population dynamics follow the same patterns. Our population has not yet reached equilibrium, but we should expect a normal population curve over time. The population of a species in an ecosystem is influenced by the availability of limited resources. When more than one species need the same limited resource (light, water, food, space, etc.), the species compete.*

6. Reducing the diversity of living things in our environment affects humans. *(T) Loss of species can eliminate species we rely on for controlling pests, for food sources, for economic benefits, for medicinal uses, and for environmental health indicators.*

GENETICS PROBLEMS

The genetics problems ask participants to explain the inheritance of dominant, recessive, codominant, and sex-linked traits and how two alleles for the same gene can lead to three different phenotypes. These problems focus on the crosscutting concept of Patterns and involve students in a number of science practices included in the *Next Generation Science Standards* (*NGSS*; NGSS Lead States 2013), most notably Developing and Using Models (science and engineering practice [SEP] 2) and Constructing Explanations and Designing Solutions (SEP 6).

Big Ideas

Genotype to Phenotype

- Genes code for proteins.
- Proteins make up the structure of cells and do the work of the cells. Sometimes these structures or the work that proteins do have visible manifestations called traits.
- A given gene may have more than one variant, each resulting in an altered protein that, in turn, gives rise to a different version of the trait.

Codominance

When an organism has two variants of a gene, the protein produced from one variant often determines (or dominates) the trait or phenotype even though the other gene variant produces a working protein. However, for some genes, the organism with two different variants has a phenotype that is either a mix of two variants or shows some features of both variants.

Sex-Linked Traits and Multigenic Traits

- For many organisms, the two sex chromosomes are different from each other. In mammals, the X chromosome has many more genes on it than the Y chromosome. Therefore, a male (XY) will have only one copy of many of the genes on the X chromosome, whereas females (XX) will have two copies.
- To keep the "dose" of genes on the X chromosome the same, in females one of the two X chromosomes is randomly inactivated early on in development. Once

one of the two X chromosomes in a cell is inactivated, all of the cells derived from that particular cell (often visible as a spot or patch of fur or skin with a particular trait) have the same active X chromosome.

- Many traits are controlled by multiple genes.

Mendelian Genetics

- In sexually reproducing organisms, each offspring gets one set of chromosomes from each parent.
- The segregation of half of the parent's chromosomes into a sex cell is a random process. Thus, one parent may pass on different combinations of chromosomes to each offspring.

Conceptual Barriers

Common Problems in Understanding

- Although many students can manipulate Punnett squares, they may not understand what the different symbols inside and outside the square represent. They may not realize that the letters outside of the square represent possible gametes, whereas those inside the square represent possible offspring. They may also be confused by the difference between the ratio of offspring phenotypes predicted by a Punnett square and the actual outcomes.
- Many students think of the X form of a duplicated chromosome whenever they hear the word chromosome. They use the same representation for unduplicated chromosomes, duplicated chromosomes, and chromosome pairs. This becomes apparent when they label different gene variants on a duplicated chromosome, as in Figure 7.1.
- Many students do not connect what they know about DNA coding for proteins with what they know about genetics.

Figure 7.1. Misrepresentation of Replicated Chromosome

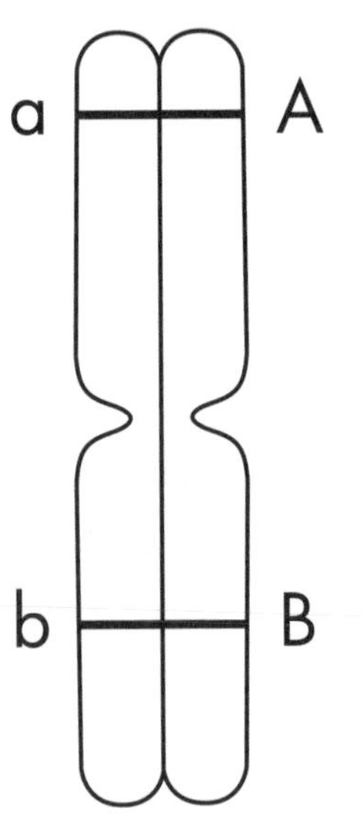

Common Misconceptions

- Recessive genes do not code for an active protein.
- Parents must have a trait or disease for the offspring to inherit that trait or disease.

Interdisciplinary Connections

A number of interdisciplinary connections can be made with the problems in this chapter. For example, connections can be made to language arts, geography, social studies, mathematics, art, and technology (see examples in Box 7.1).

Box 7.1. Sample Interdisciplinary Connections for Genetics Problems

- **Language arts:** Read about the history of the six-toed cats in Key West, Florida. Read the story "The Cat That Walked by Himself" (Kipling 1902).
- **Geography:** Explore the distribution and evolutionary history of cats around the world.
- **Social atudies:** Read articles about the role of government in regulating genetic manipulation of organisms.
- **Mathematics:** Calculate probabilities of inheriting specific traits if given parents' genotypes, or conduct chi-square analysis of actual data compared with expected outcomes.
- **Art:** Produce creative images of the family tree of the family of cats in the PBL problems.
- **Technology:** Use video or animation software to produce a short movie about inheritance in cats.

References

Kipling, R. 1902. *The cat that walked by himself and other stories.* London: British Library.

NGSS Lead States. 2013. *Next Generation Science Standards: For states, by states.* Washington, DC: National Academies Press. *www.nextgenscience.org/next-generation-science-standards.*

Genetics Problem 1: Pale Cats

Alignment With the *NGSS*

PERFORMANCE EXPECTATIONS	• *MS-LS3-1:* Develop and use a model to describe why structural changes to genes (mutations) located on chromosomes may affect proteins and may result in harmful, beneficial, or neutral effects to the structure and function of the organism. • *MS-LS3-2:* Develop and use a model to describe why asexual reproduction results in offspring with identical genetic information and sexual reproduction results in offspring with genetic variation. • *HS-LS3-2:* Make and defend a claim based on evidence that inheritable genetic variations may result from: (1) new genetic combinations through meiosis, (2) viable errors occurring during replication, and/or (3) mutations caused by environmental factors.
SCIENCE AND ENGINEERING PRACTICES	• Asking Questions and Defining Problems • Developing and Using Models • Using Mathematics and Computational Thinking • Constructing Explanations and Designing Solutions • Engaging in Argument From Evidence
DISCIPLINARY CORE IDEAS	• *LS3.A-MS Inheritance of Traits:* Genes are located in the chromosomes of cells with each chromosome pair containing two variants of each of many distinct genes. Each distinct gene chiefly controls the production of specific proteins, which in turn affects the traits of the individual. Variations of inherited traits between parent and offspring arise from genetic differences that result from the subset of chromosomes (and therefore genes) inherited. • *LS3.B-MS Variation of Traits:* In sexually reproducing organisms, each parent contributes half of the genes acquired (at random) by the offspring. Individuals have two of each chromosome and hence two alleles of each gene, one acquired from each parent. These versions may be identical or may differ from each other. • *LS1.A-HS Structure and Function:* All cells contain genetic information in the form of DNA molecules. Genes are regions in the DNA that contain the instructions that code for the formation of proteins, which carry out most of the work of cells.
CROSSCUTTING CONCEPTS	• Patterns • Cause and Effect: Mechanism and Explanation • Stability and Change

Keywords and Concepts

Genotype to phenotype, inheritance of traits

Problem Overview

A group of students wonders how a single cat can have kittens with different coat colors. Full-color versions of all the problem's images are available on the book's Extras page at *www.nsta.org/pbl-lifescience.*

GENETICS PROBLEM 1

Page 1: The Story

Pale Cats

Ms. Evan's ninth-grade biology class was visiting Kinder Kare Kitten and Kat Palace, a shelter for kittens and cats without homes. Her students were instructed to use their cell phones and cameras to take pictures of the cats and kittens housed there. They were going to upload the pictures when they got back to class and use the observed cat coat colors and patterns to explore some basic genetics. Some kittens were by themselves, and others were in family units with their mother and siblings.

"Wow—I never realized how different the kittens from one mom can be!" said Rosa. Dani said, "Look! Here's a pale mom that has three bright orange kittens, plus a patchy black-and-orange one. I'm going to take a picture of these guys." (See Figure 7.2.) Her friend, Sasha, was looking at a gray cat. "Oh, poor gray kitty! It didn't get very much color. Maybe it was the runt of the litter." "Maybe it started out black, but didn't get enough to eat," replied Dani.

Figure 7.2. Dark and Pale Orange

Note: A full-color version of this figure is available on the book's Extras page at *www.nsta.org/pbl-lifescience.*

Ms. Evan heard them talking and said, "Nothing has happened to the gray kittens. The difference between the dark orange and pale orange cats is their genes. Remember our genetics mantra: 'DNA to protein to phenotype'? The difference between the dark orange and the pale orange kittens is a great example of this mantra. Do you see the black and gray kittens over there? The same explanation can be used for these two." Rosa said, "The same explanation? Is a gray kitten kind of like adding white paint to black? That makes sense to me." Dani asked, "Does that mean that we'll get to use our picture of these kittens in class? Cool!"

Your Challenge: *Explain the difference between cats with pale versus dark coats (gray versus black and light orange versus dark orange) in terms of DNA, proteins, and traits. Explain the inheritance of pale coats.*

GENETICS PROBLEM 1

Page 2: More Information

Pale Cats

Ms. Evan's class had already looked up *phenotype* in the dictionary and found that it was defined as the set of observable characteristics of an individual resulting from the interaction of its genotype with the environment. As a class, they had decided that the important part of the definition was *observable characteristics*. They also knew that an organism's phenotype is a result of its genes. Figure 7.3 shows black and grey cats.

When they got back to class, Ms. Evan gave them this information:

- The darkness of cat coats depends on the expression of one gene, *dilute*.
- Pale cats have less pigment in each hair.
- Several gene products are required for pigment to end up in a cat hair. An altered form of one of these proteins produces pale (dilute) cats.
- Pale cats can be born to parents that both have dark coats.

Figure 7.3. Black and Gray Cats

Note: A full-color version of this figure is available on the book's Extras page at *www.nsta.org/pbl-lifescience*.

Your Challenge: *Explain the difference between cats with pale versus dark coats (gray versus black and light orange versus dark orange) in terms of DNA, proteins, and traits. Explain the inheritance of pale coats.*

GENETICS PROBLEM 1

Teacher Guide

Pale Cats

Problem Context

Cats are a species common in students' homes, and they exhibit a wide variety of coat colors. These coat colors are examples of common patterns of inheritance, so cats make a good model for learning how to predict the appearance of offspring and explain observed results. One advantage in using cats is that there is much research published showing a few genes that serve as examples of autosomal and sex-linked traits. The calico color also demonstrates the deactivation of one of the X chromosomes in females. In this problem, the trait represents a simple dominance.

Related Contexts

Simple dominance can also be shown in coat colors of mice, colors of many types of flowers, and human traits such as tongue rolling.

> *Model response:* Pale cats differ from cats with dark coats in one gene, the dilute gene. The most common variant of the dilute gene codes for a protein that participates in packing a lot of pigment into a hair. As long as there is some of the common form of the protein present, lots of pigment will be packed into the hairs. If only protein made from the dilute variant of the gene is present, less pigment will be packed into the hairs, resulting in a cat with pale (or dilute) color. The color of the pigment that is packed into the hairs is determined by other genes.
>
> A pale coat is a recessive trait, meaning that an animal must have two copies of the genes for that trait; otherwise, the trait is not visible. Parents with dark coats may have one gene for paleness. If both parents happen to pass on the dilute variant to a kitten, the kitten's coat will be pale.

Activity Guide

Students can answer the challenge by putting together the information provided. Encourage them to do this rather than simply finding information about the dilute gene online. If available, students can augment their experience by looking at hairs from dark- and light-color cats with a magnifying lens. Putting the hairs on a black background helps make the difference in the hairs more visible.

GENETICS PROBLEM 1

Assessment

Pale Cats

Transfer Task

Give students pictures of light and dark cats. Have them predict whether different combinations of parents could have pale (dilute) kittens. To do this, they should predict which versions of the dilute gene the parents are carrying. There may be more than one possibility (see Figure 7.4).

Model response: Any combination of adult cats could possibly have dilute kittens if those with dark coats are carrying one copy of the dilute variant. Any parents that have pale coats have two copies of the dilute variant of the gene [SEP 6: Constructing Explanations and Designing Solutions; CC 1: Patterns].

Figure 7.4. How to Get Pale Kittens

Parents:	pale x pale	**Parents:**	pale x dark	**Parents:**	dark x dark
Genes:	dd x dd	**Genes:**	dd x D?	**Genes:**	D? x D?
Kittens:	all dd (pale)	**Kittens:**	Dd dark dd pale	**Kittens:**	DD dark Dd dark dd pale
		Only if	? = d	**Only if**	? = d for both parents

General Question

Explain what is meant by "DNA to protein to phenotype."

Model response: Segments of DNA called *genes code* for proteins. Proteins make up the structure of cells and do the work of the cells. Sometimes these structures or the work that proteins do have visible manifestations called *traits*. A given gene may have more than one variant, each resulting in an altered protein that, in turn, gives rise to a different version of the trait.

Application Questions

APPLICATION QUESTION 1

Gregor Mendel, considered to be the founder of modern genetics, found that sometimes when he cross-pollinated (mated) two pea plants with red flowers, some of their offspring had white flowers. Explain these results in terms of the DNA, protein, and phenotype.

> *Model response:* Pea plants probably have a gene (a segment of DNA) that codes for a protein that affects either the production of red flower pigment or the packing of red pigment into petals. As long as some of the usual form of the protein is present, red pigment is in the petals. If only the variant form of the protein is present, no red pigment will be in the flowers and they will be white. In the cases where two red-flowering plants gave rise to white-flowering plants, each of the parental plants must have been carrying one copy of the gene variant associated with white flowers. Some of their offspring inherited that gene variant from both parents and therefore had white flowers. However, the offspring from those parents were more likely to inherit the gene variant associated with red flowers from at least one of the parental plants. This means that the resulting plants are more likely to have red flowers than white flowers [SEP 6: Constructing Explanations; CC 1: Patterns].

APPLICATION QUESTION 2

At 5 years old, Maura was extremely sun sensitive. She freckled extensively (both in numbers and darkness of freckles), and her eyes hurt whenever she went out in the sun. No one else in the family shared these characteristics. Genetic testing showed that Maura had *xeroderma pigmentosum*, which interferes with the normal repair of sun-damaged DNA. Figure 7.5 presents a pedigree showing the genetic data from Maura's family. Describe the patterns you see in the pedigree data. Explain whether the dilute trait in cats would have a pedigree similar to Maura's.

Figure 7.5. Maura's Pedigree

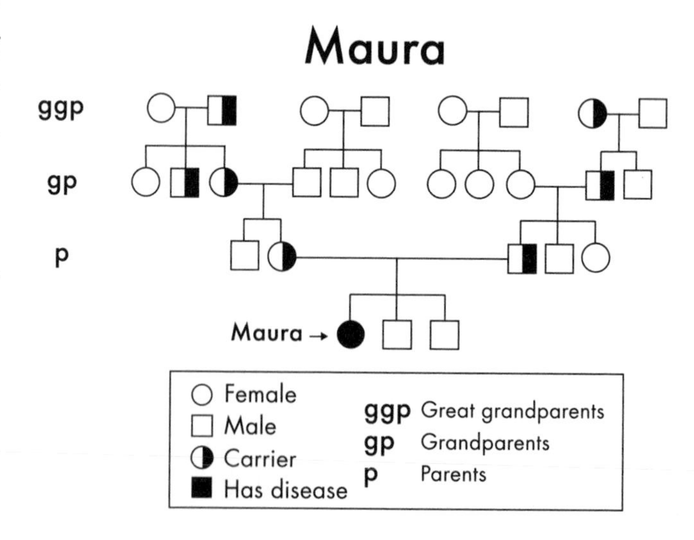

Model response: Maura's pedigree shows several consistent patterns [CC 1: Patterns]. Carriers (people with one, but not two, of the disease-causing variant) have one parent who is also a carrier. Both males and females can be carriers. No one with only one parent who is a carrier has the disease [SEP 4: Analyzing and Interpreting Data]. These patterns are the same as what you would expect for the dilute trait in cats. No kittens that have only one parent carrying the dilute variant are born with pale fur.

Common Beliefs

Indicate whether the statements are true (T) or false (F), and explain why you think so (*model responses shown in italics*).

1. Dominant gene variants prevent recessive gene variants from producing protein. *(F) This is rarely the case. Production of protein from one gene rarely affects the expression (use of) the other gene. The two genes are usually used independently. It should be noted that recessive gene variants often produce proteins that are somewhat functional. If this were not the case, individuals with two copies of the recessive gene would not have any functioning protein.*

2. The dominant gene variants are more common than the recessive variants. *(F) A dominant trait may be less common, even very rare. The percentage of a population that has the trait depends on how many copies of the gene with that variant are in the population. Dominance is not the result of frequency, but instead it describes which trait shows up in an individual with one copy of each variation of the trait.*

3. If a parent passes on a specific trait to one offspring, the chances of that trait being passed on to the next offspring is reduced. *(F) Each sperm or egg cell is produced from a separate meiotic cell division, so the chance of passing on a given allele is identical for each offspring.*

Genetics Problem 2: Black and White and Spots All Over

Alignment With the *NGSS*

PERFORMANCE EXPECTATIONS	• *MS-LS3-2:* Develop and use a model to describe why asexual reproduction results in offspring with identical genetic information and sexual reproduction results in offspring with genetic variation. • *HS-LS3-2:* Make and defend a claim based on evidence that inheritable genetic variations may result from: (1) new genetic combinations through meiosis, (2) viable errors occurring during replication, and/or (3) mutations caused by environmental factors.
SCIENCE AND ENGINEERING PRACTICES	• Asking Questions and Defining Problems • Developing and Using Models • Using Mathematics and Computational Thinking • Constructing Explanations and Designing Solutions • Engaging in Argument From Evidence
DISCIPLINARY CORE IDEA	• *LS3.B-MS Variation of Traits:* In sexually reproducing organisms, each parent contributes half of the genes acquired (at random) by the offspring. Individuals have two of each chromosome and hence two alleles of each gene, one acquired from each parent. These versions may be identical or may differ from each other.
CROSSCUTTING CONCEPTS	• Patterns • Cause and Effect: Mechanism and Explanation • Stability and Change

Keywords and Concepts

Codominance, inheritance of traits

Problem Overview

Students explore how spotted coat color is inherited in cats. Full-color versions of all the problem's images are available on the book's Extras page at *www.nsta.org/pbl-lifescience.*

GENETICS PROBLEM 2

Page 1: The Story

Black and White and Spots All Over

Ms. Evan's students were going through their class collection of photos of cats. Max wanted to find a photo of an all-white cat, which he would call Ghost or Cloud. Finally, he saw one that might work. "Hey, look at this one—he is mostly white with a few black spots on his head, back and tail," said Max (Figure 7.6). "Yeah, and look at mine—she is almost the reverse, mostly black with a few white spots," said Minnie. "And here is one that is totally black, with no white anywhere!" (See Figure 7.7.)

Ms. Evan said, "These cat coat patterns are a great example of how the variants of a single gene can produce three different phenotypes. In this case, the spotting gene produces the different coat colors."

Figure 7.6. Max's Cat Picture

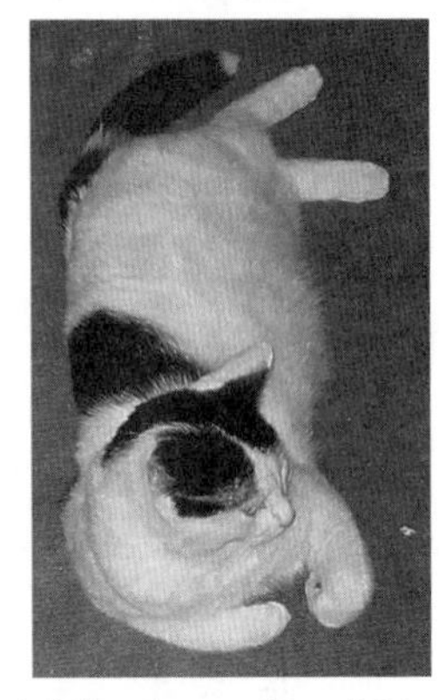

Note: A full-color version of this figure is available on the book's Extras page at *www.nsta.org/pbl-lifescience.*

Your Challenge: *Explain in terms of DNA and proteins how the two variants of the spotting gene can produce three phenotypes.*

Figure 7.7. Minnie's Cat Pictures

Note: A full-color version of this figure is available on the book's Extras page at *www.nsta.org/pbl-lifescience.*

GENETICS PROBLEM 2

Page 2: More Information

Black and White and Spots All Over

Snapdragons have a gene whose protein determines the color of the pigment in the flowers. Snapdragon plants that have two copies of one gene variant have red flowers. Plants with two copies of the other gene variant have white flowers. Plants with one copy of each gene variant make both red and white pigment, and their flowers are pink. Figure 7.8 shows snapdragon variants.

Figure 7.8. Snapdragon Variants

Note: A full-color version of this figure is available on the book's Extras page at *www.nsta.org/pbl-lifescience*.

Bell peppers have a gene that codes for a protein that determines the amount of pigment that ends up in the fruit (Figure 7.9). Red peppers develop in plants that have two copies of the gene variant associated with lots of red pigment. Pale red peppers are pale because they don't have very much pigment. They develop on plants that have one copy of the gene variant associated with lots of pigment and one copy of the gene variant associated with small amounts of pigment. If a plant has two copies of that gene variant, it puts essentially no red pigment into its fruits, and the fruits remain green.

Figure 7.9. Bell Pepper Variants

Note: A full-color version of this figure is available on the book's Extras page at *www.nsta.org/pbl-lifescience.*

Your Challenge: *Explain in terms of DNA and proteins how the two variants of the spotting gene can produce three phenotypes.*

GENETICS PROBLEM 2

Teacher Guide

Black and White and Spots All Over

Problem Context

Cats are a species common in students' homes, and they exhibit a wide variety of coat colors. These coat colors are examples of common patterns of inheritance, so cats make a good model for learning how to predict the appearance of offspring and explain observed results. One advantage in using cats is that there is much research published showing a few genes that serve as examples of autosomal and sex-linked traits. Spotting is an example of incomplete dominance, sometimes referred to as "blending."

Related Contexts

Incomplete dominance can also be shown in coat color of rabbits (black × white produces gray). In humans, a similar trait is hair texture (straight, wavy, or curly).

> *Model response:* The gene for spotting codes for a protein that affects the amount of white in a cat's coat. One variant of the spotting gene is associated with absence of white fur. The other variant of the spotting gene is associated with white fur (covering more than half of the cat's body). A cat with one copy of each gene variant has an intermediate amount of white fur (covering less than half of its body).

Activity Guide

The goal of this problem is to figure out the gene for spotting in cats by applying what students learn from the snapdragon and bell pepper cases. Students will have the least trouble figuring out this problem if they focus on trying to understand the proteins being made in each situation. Making more or less of a pigment protein depending on gene dosage is a straightforward concept, if students let go of the oft-taught dominant/recessive scenario. For this reason, it may be helpful as facilitator to review the challenge from time to time with its requirement that the explanation include both DNA (genes) and proteins.

GENETICS PROBLEM 2

Assessment

Black and White and Spots All Over

Application Questions

APPLICATION QUESTION 1

A blond woman of northern European descent is married to a Chinese man with black hair. They have three sons. Predict the color of the sons' hair. For the purpose of this question, you may assume that human hair color is controlled by one gene. (In reality, human hair color is more complex than this, because it is controlled by more than one gene.) Justify your prediction by explaining what you think occurred with the genes, protein, and hair color. Your prediction should include any assumptions about how the gene products interact.

Model response: I predict that the Chinese father has two copies of the gene variant associated with black hair. Perhaps this gene variant codes for a protein that adds dark pigment to lighter-color hair. The blond mother might have two copies of the gene variant associated with blond hair. Perhaps this gene variant codes for a version of the protein that adds much less pigment to the hair. If I assume that black hair is a dominant trait, I predict that all of the sons will have black hair. They will have one copy of each gene variant. The one copy associated with black hair will produce enough protein that adds dark pigment to lighter hair so that their hair will appear dark. If I assume that hair color is not a dominant trait, I predict that the sons will all have brown hair. They will produce a mix of the two protein variants that will add an intermediate amount of pigment to the hair.

Another possibility: Perhaps the gene variant for black hair codes for a protein that participates in making black hair pigment. The gene variant for blond hair codes for a protein that participates in making yellow pigment. If this is the case, the sons will have brown hair that is a mix of black and yellow pigment [SEP 5: Using Mathematics and Computational Thinking; SEP 6: Constructing Explanations and Designing Solutions; CC 1: Patterns].

APPLICATION QUESTION 2

It is very rare for a Chinese couple to have children with any hair color other than black. What does this tell you about the genetics of Chinese black hair? Explain your reasoning.

Model response: Chinese people probably have two of the same variants of the hair color gene. If they carried a gene variant associated with light hair, it would not be uncommon for two carriers to occasionally (chance: 1 in 4) have a blond child [SEP 5: Using Mathematics and Computational Thinking; SEP 6: Constructing Explanations and Designing Solutions].

APPLICATION QUESTION 3

Describe the patterns you would expect to see in the inheritance of spots. For example, could a cat that had a small amount of white have a kitten with a lot of white? Use your list of patterns to draw a cat pedigree that has at least three generations, including at least one cat with no white, one with white on less than half its body, and one with white on more than half its body.

Figure 7.10. Possible Pedigree for a Cat With Spots

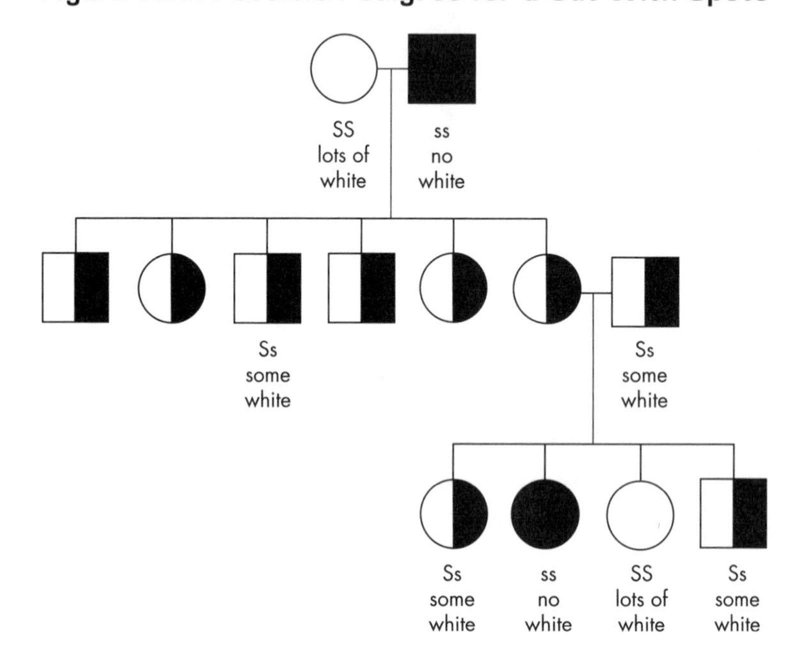

Model Response: Cats with no white can have any combination of parents with no white and some white. They cannot have two parents with lots of white. Cats with some white cannot have two parents without white, nor can they have two parents with lots of white. Here are possible ways to get a cat with some white:

- No white × some white or lots of white
- Some white × any kind of cat
- Lots of white × some white or no white

Cats with lots of white cannot have two parents with no white. Note that not all of the kittens from these crosses will have the desired trait. These are crosses where the desired trait is a possibility. Figure 7.10 shows a possible pedigree [CC 1: Patterns].

Genetics Problem 3: Calico Cats

Alignment With the *NGSS*

PERFORMANCE EXPECTATIONS	• *MS-LS3-2:* Develop and use a model to describe why asexual reproduction results in offspring with identical genetic information and sexual reproduction results in offspring with genetic variation. • *HS-LS3-2:* Make and defend a claim based on evidence that inheritable genetic variations may result from: (1) new genetic combinations through meiosis, (2) viable errors occurring during replication, and/or (3) mutations caused by environmental factors.
SCIENCE AND ENGINEERING PRACTICES	• Asking Questions and Defining Problems • Developing and Using Models • Using Mathematics and Computational Thinking • Constructing Explanations and Designing Solutions • Engaging in Argument From Evidence
DISCIPLINARY CORE IDEA	• *LS3.B-MS Variation of Traits:* In sexually reproducing organisms, each parent contributes half of the genes acquired (at random) by the offspring. Individuals have two of each chromosome and hence two alleles of each gene, one acquired from each parent. These versions may be identical or may differ from each other.
CROSSCUTTING CONCEPTS	• Patterns • Cause and Effect: Mechanism and Explanation • Stability and Change

Keywords and Concepts

Sex-linked and multigenic traits

Problem Overview

A group of students tries to explain the genetic cause of calico and tortoiseshell coat color in cats. Full-color versions of all the problem's images are available on the book's Extras page at *www.nsta.org/pbl-lifescience.*

GENETICS PROBLEM 3

Page 1: The Story

Calico Cats

Lena was interested in calico cats, with their orange spots and their dark spots on a white background (Figure 7.11). When she was little, her grandfather often recited for her Eugene Field's poem called "The Duel" about a fight between a gingham dog and a calico cat (you can read the poem at *www.poets.org/poetsorg/poem/duel*). He also had a very large calico cat named Gingham.

Lena collected a bunch of pictures of different calico cats from Ms. Evan's class collection. Some had a lot of white, and some had very little. Some had large spots, and others had smaller spots. Some had black spots, and some had dark spots that looked like broken-line stripes. But they all had the three colors—white background, orange spots, and dark spots.

Ms. Evan said, "The calico pattern is the result of the expression of three genes. You already know about one—the gene for spotting. You need to figure out the genetics behind the black and orange patches. Remember that we saw a cat with a very sharp line down its nose dividing an orange patch from a black patch. Orange and black patchy cats don't always have white on them. The ones without white are called tortoiseshell cats. And here's a hint: Calico cats and tortoiseshell cats (orange and black cats) are almost always females. That means that you need to learn about a gene found on the X chromosome."

Figure 7.11. Calico Cats

Note: A full-color version of this figure is available on the book's Extras page at *www.nsta.org/pbl-lifescience*.

Your Challenge: *Explain the genetics of calico cats. This means that you need to explain the inheritance of three genes—spotting, black, and orange—and explain why male cats are rarely calico.*

GENETICS PROBLEM 3

Page 2: More Information

Calico Cats

Here is some of the information that Ms. Evan's class found:

- Orange cats are more likely to be male.
- Female cats have two X chromosomes; male cats have one X and one Y chromosome. The Y chromosome is small and does not have as many genes on it as the X chromosome. This means that for many genes on the X chromosome, females have two copies, but males have only one.
- To make sure that males and females have the same amount of the proteins made from those genes on the X chromosome, cells in females use only one of their two X chromosomes. Early in mammalian development, one X chromosome is chosen at random and permanently inactivated in each non-sex cell in a female embryo. All of the cells that develop from that cell will have the same X chromosome inactivated.

Your Challenge: *Explain the genetics of calico cats. This means that you need to explain the inheritance of three genes—spotting, black, and orange—and explain why male cats are rarely calico.*

GENETICS PROBLEM 3

Teacher Guide

Calico Cats

Problem Context

Cats are a species common in students' homes, and they exhibit a wide variety of coat colors. These coat colors are examples of common patterns of inheritance, so cats make a good model for learning how to predict the appearance of offspring and explain observed results. One advantage in using cats is that there is much research published explaining the genes that serve as examples of autosomal and sex-linked traits. The calico color demonstrates both a sex-linked trait and the deactivation of one of the X chromosomes in females.

Related Context

In humans, sex-linked traits include red-green colorblindness and hemophilia. These traits can be used to demonstrate this concept, although having actual individuals from many generations can be hard to find.

> *Model response:* Calico cats are usually female cats. They have white fur because they have one or two copies of the spotting gene variant associated with white fur [see Genetics Problem 2]. Because they are usually female cats, calico cats have two X chromosomes (males have one X and one Y chromosome). One X chromosome has a gene variant coding for a protein that turns black hair pigment into orange pigment. The other X chromosome has a variant of the gene that leaves the pigment black. Like males, female cats need only one X chromosome, so as they develop they inactivate one or the other of their X chromosomes. The choice of which one to inactivate is a random event. Patches of skin that come from cells where the orange variant was inactivated and the black variant is active will produce black fur. The orange patches come from cells where the black variant was inactivated and the orange variant is active (Figure 7.12). Male cats can't be calicos because they have only one X chromosome. If that chromosome has the black gene variant, their whole coat will be the usual black. If the X chromosome has the orange gene variant, their whole coat will be orange.
>
> Tortoiseshell cats also are female cats with one copy each of the black and orange gene variants. Only one X chromosome is active in each patch of skin and the gene variant on the active X chromosome determines the fur color. Tortoiseshell cats do not have any white fur because they don't have the spotting gene variant.

Figure 7.12. Calico Cat Genetics

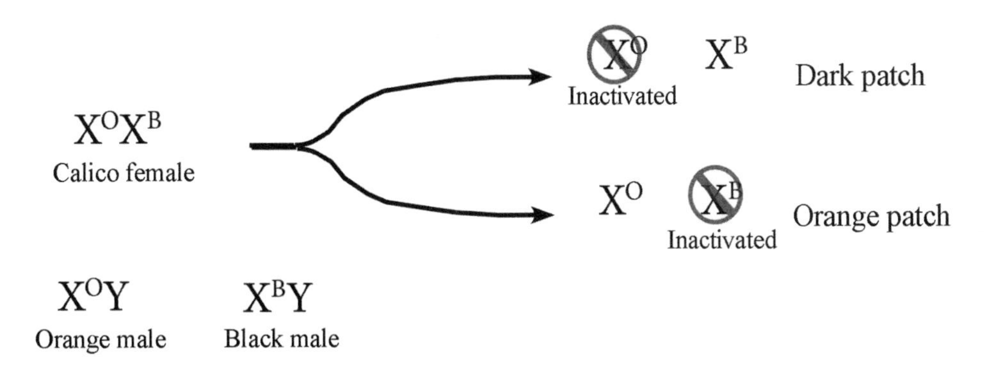

GENETICS PROBLEM 3

Assessment

Calico Cats

Transfer Task

Queen Victoria of Great Britain married her cousin, Prince Albert of Saxe-Coburg and Gotha. They had nine children. One of their sons had hemophilia. Hemophilia is a genetic disease in which the blood cannot coagulate (clot). This leads to extensive bleeding even with the smallest of injuries.

The pedigree in Figure 7.13 shows the inheritance of this disease in four generations of Victoria's family. For simplicity, Victoria's first two children, Princess Victoria and Prince Albert (later King Edward VII), are not included because no sign of hemophilia has been seen in these two families. Also note that four of Victoria's nine children did not marry or have children. This leaves three of Victoria's children whose families have members with hemophilia.

Figure 7.13. Queen Victoria's Pedigree

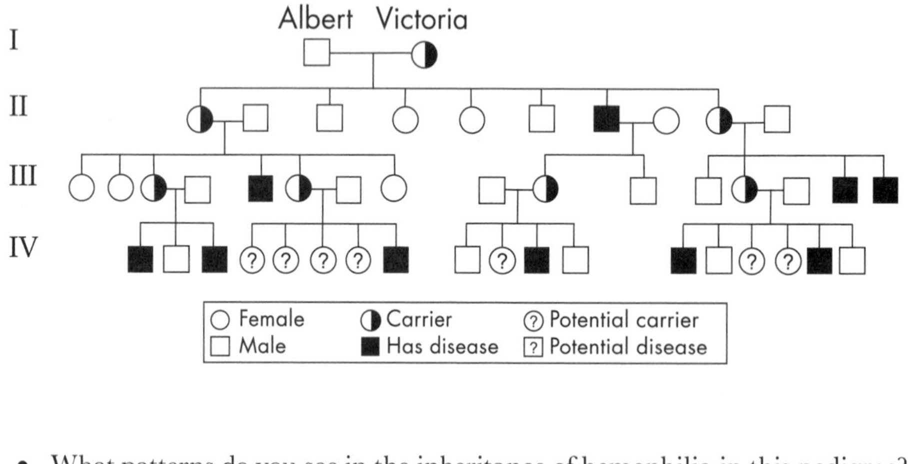

- What patterns do you see in the inheritance of hemophilia in this pedigree?
- Which coat color gene in cats has similar patterns of inheritance? Explain your answer.

- A number of the women in the fourth generation who are identified with question marks are royalty, or at least aristocracy, of Europe. Calculate the chances that these women are carriers for hemophilia.

Model response:

1. Only males get the disease. Females are carriers. Males who have the disease have mothers who are carriers and fathers who don't have the disease [CC 1: Patterns].
2. This is like the gene for orange coat color in cats. The genes for orange fur and for hemophilia are carried on the X chromosome. To have hemophilia, a woman must inherit an X chromosome with the gene variant for hemophilia from both parents. That means that hemophilic women have hemophilic fathers, and their fathers must live long enough with the disease to have children. That is why hemophilia is very rare in women. Orange female cats have orange fathers, but it is not rare that the fathers survive to have children.
3. None of the women with question marks are likely to have hemophilia because their fathers do not have hemophilia. Therefore, the X chromosome they inherited from their father did not have the hemophilia gene variant. Their mothers are carriers, meaning that one of their X chromosomes has the hemophilia gene variant, but the other one doesn't. The "normal" gene variant on one X chromosome is enough to ensure normal blood clotting. The daughters have a 50% chance of inheriting the X chromosome with the hemophilia gene variant from their mothers and being carriers [SEP 5: Using Mathematics and Computational Thinking].

Genetics Problem 4: Cat Puzzles

Alignment With the *NGSS*

PERFORMANCE EXPECTATIONS	• *MS-LS3-1:* Develop and use a model to describe why structural changes to genes (mutations) located on chromosomes may affect proteins and may result in harmful, beneficial, or neutral effects to the structure and function of the organism. • *MS-LS3-2:* Develop and use a model to describe why asexual reproduction results in offspring with identical genetic information and sexual reproduction results in offspring with genetic variation.
SCIENCE AND ENGINEERING PRACTICES	• Asking Questions and Defining Problems • Developing and Using Models • Using Mathematics and Computational Thinking • Constructing Explanations and Designing Solutions
DISCIPLINARY CORE IDEAS	• *LS3.B-MS Variation of Traits:* In sexually reproducing organisms, each parent contributes half of the genes acquired (at random) by the offspring. Individuals have two of each chromosome and hence two alleles of each gene, one acquired from each parent. These versions may be identical or may differ from each other. • *LS1.A-HS Structure and Function:* All cells contain genetic information in the form of DNA molecules. Genes are regions in the DNA that contain the instructions that code for the formation of proteins, which carry out most of the work of cells.
CROSSCUTTING CONCEPTS	• Patterns • Cause and Effect: Mechanism and Explanation • Stability and Change

Keywords and Concepts

Patterns of inheritance, genotype to phenotype, codominance, sex-linked and multigenic traits

Problem Overview

Students are asked to determine if all the different-color kittens in a litter could be related to the same father. Full-color versions of all the problem's images are available on the book's Extras page at *www.nsta.org/pbl-lifescience.*

GENETICS PROBLEM 4

Page 1: The Story

Cat Puzzles

Ms. Evan's class was startled to learn that the kittens in one litter born at the same time to one mother might have different fathers. This happens when a female cat mates with more than one male. The class wondered if they could tell from the variety of kittens if there had been more than one father. They chose pictures of two mother cats with many kittens of various colors. One mother they named Blossom. Blossom was pale orange with some white on her. She had four kittens: Eanie and Meanie were bright orange with a little white, Mo was orange without white, and Robin Hood was a tortoiseshell (Figure 7.14). The other mother they named Carmen. She was a pale orange cat with four kittens: Penny was a tortoiseshell, Matt and Matilda were both orange (but Matt didn't have stripes and Matilda did), and Midnight was all black without stripes (Figure 7.15). The class did not worry about the gender of the kittens when they named them.

Your Challenge: *Use cat coat color genetics to determine if all of the kittens in Blossom's or Carmen's litters could have had the same father.*

Figure 7.14. Blossom and Kittens

Figure 7.15. Carmen and Kittens

Note: Full-color versions of the figures above are available on the book's Extras page at *www.nsta.org/pbl-lifescience*.

GENETICS PROBLEM 4

Page 2: More Information

Cat Puzzles

Ms. Evan's class put together a guide to cat coat color genetics. The guide is below.

Cat Coat Color Genetics

Orange: sex-linked; orange pigment (orange) is produced instead of the black pigment (*Note:* X^o is the orange allele on the X chromosome, and X^+ is the "not orange" allele on the X chromosome.)

- X^oY—orange male
- X^+Y—black male
- X^oX^o—orange female
- X^+X^o—tortoiseshell cat (blotches of orange and tabby/black)

Spotting: incomplete dominance

- SS—white spots on more than half of body
- Ss—white spots on less than half of body
- ss—no white spots

Dilute: codes for packing or distribution of pigment in hairs

- DD or Dd—normal amount of pigment in each hair
- dd—less pigment packed into each hair

Agouti: results in individual hair striping; to have striped or swirly coats, cats must have agouti hairs. Dominant alleles produce hairs that are dark with light bands. The amount of striping may be variable, giving lighter or darker tabbies (see tabby section).

- AA or Aa—agouti hairs
- aa—solid pigmented hairs

Black: codes for pigment

- BB or Bb—black pigment (or orange if combined with X^o)
- bb—brown pigment

Tabby: Many variants of this gene determine the pattern of striping. Tabby patterns can be dark lines coming down from the spine or thinly striped flanks, stripes all over, or whorls and blotches rather than stripes. If agouti is present, tabby patterns can be seen in most colors of cats.

White: removes all melanin from coat; reduces pigment in eyes, resulting in blue or yellow eyes; often causes deafness, especially with blue eyes. This trait represents a good example of both *epistasis* (how one gene can regulate other genes) and *pleiotropy* (how one gene can influence more than one trait).

- WW or Ww—white
- ww—no white

Your Challenge: *Use cat coat color genetics to determine if all of the kittens in Blossom's or Carmen's litters could have had the same father.*

GENETICS PROBLEM 4

Teacher Guide

Cat Puzzles

Problem Context

The set of traits seen in cat coat color will allow students to practice analyzing pedigrees that show traits within a family. By looking at the characteristics of offspring and using patterns of inheritance (CC 1: Patterns), students can infer the genotypes of parents.

Related Contexts

Many pedigree activities use animals (e.g., horses, dogs, and cattle) or humans. Genetic counselors often start with family pedigrees for disorders such as hemophilia, cystic fibrosis, sickle cell anemia, and many more traits.

> *Model response:* If a cat has a dominant trait, you know that at least one of the two genes for that trait is the variant for that dominant trait, but you don't know whether the other gene codes for the dominant or recessive variant, unless one of the parents has only recessive variants. (This is why Eanie and Meanie must have at least one recessive variant of the dilute gene.) If a cat has a recessive trait, you know that both of the genes for that trait are the recessive variant.
>
> We know Blossom's genotype. The amount of white on her body indicates that she has one dominant and one recessive variant of the spotting gene (Ss). Because she is pale, the only possibility for the dilute gene is that she has two recessive gene variants (ss). She is an orange female and not a calico, which means that she has orange variants on both X chromosomes (X^oX^o).
>
> As shown in the data in Table 7.1, all of Blossom's kittens could have had the same father if Eanie and Meanie are male. The father's genotype was Ss or ss, DD or Dd (the former is more likely given that all four kittens were not dilute), X^+Y. An orange male (X^oY) could not have fathered Robin Hood (who is a female). A cat with lots of white (SS) could not have fathered Mo.
>
> We know Carmen's genotype. She is pale colored, so she must have two copies of the dilute gene variant (dd). She is an orange female, so her X chromosomes must both have the orange gene variant (X^oX^o), otherwise she would have tortoiseshell coloring (X^oX^+). Because she is striped, she must have at least one dominant agouti

gene variant (A). She has two kittens that are not striped. Therefore, she must have one recessive variant of the agouti gene, which she passed on to them (s).

Carmen's kittens could have all had the same father (Table 7.2, p. 182). His genotype was likely DD, because all the kittens are dark, but Carmen is pale. He was black (X^+Y), not orange, because Penny inherited a standard X from him that contributed to her tortoiseshell coat. He had at least one non-agouti gene (?a) that he passed on to the kittens without stripes.

Table 7.1. Traits and Genotypes of Cats in the Cat Puzzles Problem

NAME OF CAT	PHENOTYPE	POSSIBLE GENOTYPES
Blossom	Some spotting Dilute Orange	Ss dd X^oX^o
Mo	No white Dilute Orange	ss dd X^oX^o or X^oY
Eanie	Some spotting Dark Orange	Ss Dd X^oX^o or X^oY
Meanie	Some spotting Dark Orange	Ss Dd X^oX^o or X^oY
Robin Hood	Some spotting Dark Orange and black	Ss Dd X^oX^+

Table 7.2. More Traits and Genotypes in the Cat Puzzles Problem

NAME OF CAT	PHENOTYPE	POSSIBLE GENOTYPES
Carmen	Pale Orange Striped	dd X^oX^o Aa
Penny	Dark Tortoiseshell Not striped	Dd X^oX^+ aa
Matt	Dark Orange Not striped	Dd X^oX^o or X^oY aa
Matilda	Dark Orange Striped	Dd X^oX^o or X^oY A?
Midnight	Dark Black Not striped	Dd X^+X^+ or X^+Y aa

Activity Guide

If your students have done Genetics Problems 1–3, this problem requires very little setup. If they have not done the previous problems, we suggest having students practice using genetics by determining the genotypes of individual cats (from pictures or their own pets). Using the guide to cat coat color genetics included on Page 2 of the story in this problem, students should be able to determine the possible genotypes of almost any domestic shorthair cat. However, these genes do not apply to special breeds such as Siamese cats.

GENETICS PROBLEM 4

Assessment

Cat Puzzles

Once students are familiar with the cat coat color genes, you can use photos of cats to generate any number of assessment questions. All of these require students to use patterns of inheritance (CC 1: Patterns) to construct explanations (SEP 6).

Application Questions

APPLICATION QUESTION 1: PREDICT GENOTYPES

What are the possible genotypes of a striped black cat with no white? Don't worry about the tabby gene.

> *Model response:* A (A or a) allows for striping; B (B or b) produces black rather than brown; X^+ (X^+ or Y) means the kitten will not be orange; D (D or d) leads to dark color rather than pale; ss causes no white patches.

APPLICATION QUESTION 2: PREDICT OFFSPRING GENOTYPES AND PHENOTYPES

Martin has a solid black female cat (no striping). His friend Tarisha has a solid orange male cat (no striping). They are interested in the possibility of letting the two cats mate and selling orange and black cats at Halloween. However, they're not sure what the chances are that the kittens will be orange or black. Help them out by predicting the chances of having black or orange kittens. Because brown is very rare, assume that neither cat is carrying a gene variant for brown rather than black fur.

> *Model response:* As shown in Figure 7.16, half of the kittens will be tortoiseshell females, half will be black males.

Figure 7.16. Matrix Predicting Traits of Martin's and Tarisha's Cats

	a X^+	**a X^+**
a X^o	aa X^+ X^o	aa X^+ X^o
a Y	aa X^+ Y	aa X^+ Y

APPLICATION QUESTION 3: PREDICT GENOTYPE AND PHENOTYPE OF PARENTS

Elijah's neighbors gave him one of their cat's kittens. The kitten was a black striped tiger male with white feet. Elijah named the kitten Bart, and he often wondered about who Bart's father was. The only cat that he saw in the neighborhood besides Bart's mother was a pale orange tomcat. Could a cat that looked so different be Bart's father?

> *Model response:* A pale orange male cat would have a genotype dd (pale), $X^{o}Y$ (orange). Because the dilute allele or gene variant is recessive, this cat could have dark-color offspring. Male offspring of this cat would inherit the Y chromosome from this cat, which would not contribute to an orange coat. So the pale orange tomcat could be Bart's father [SEP 6: Constructing Explanations and Designing Solutions; CC 1: Patterns].

CELLULAR METABOLISM PROBLEMS

The cellular metabolism problems ask participants to compare the inputs, outputs, and energy transformations associated with combustion, respiration, and photosynthesis. These processes are contrasted with digestion and biosynthesis. The problems focus on the crosscutting concept (CC) Energy and Matter: Flows, Cycles, and Conservation and involve students in a number of science practices included in the *Next Generation Science Standards* (*NGSS*; NGSS Lead States 2013), most notably Developing and Using Models (science and engineering practice [SEP] 2) and Analyzing and Interpreting Data (SEP 4).

Big Ideas

Combustion and Cellular Respiration

- When sugar is burned (combustion) or metabolized (cellular respiration), it reacts with oxygen to form carbon dioxide (CO_2) and water.
- During combustion and cellular respiration, the following high-energy bonds between carbon (C), hydrogen (H), and oxygen (O) are broken: C–C, C–H, and O–O. (Breaking bonds requires energy, but these bonds are *high energy* because they are not very stable and require little energy to break.)
- More stable C–O and H–O bonds are formed, releasing more energy than was needed to break the original bonds.
- The net effect is release of energy. In the case of combustion, the energy is released as heat and light. In the case of cellular respiration, most of the energy is captured as chemical potential energy in adenosine triphosphate (ATP), which is used to do the work of the cell. Eventually, ATP ends up as heat.

Photosynthesis

- Photosynthesis is the "reverse" of cellular respiration. Sugars and molecular oxygen are made from CO_2 and water.
- Stable C–O and O–H bonds in CO_2 and water are broken. This requires a lot of energy, which ultimately comes from sunlight.

- High-energy C–C, C–H, and O–O bonds are produced. Thus, the products of photosynthesis have a lot of chemical potential energy. One of the products is sugar, which acts as food for the plant that can be used as an energy source or as building blocks.

Photosynthesis, Cellular Respiration, Digestion, and Biosynthesis

- All living things need a source of fixed (reduced/organic) carbon to make the building blocks of their bodies.
- All living things need an energy source so that their cells can do their work (chemical reactions, movement, reproduction).
- Plants (and most autotrophs) use photosynthesis to produce their own food (sugars), which serves both as the carbon source for the production of other molecules (biosynthesis) and as an energy source. (Autotrophs make their own food, whereas heterotrophs consume other material for food.)
- Animals (and most heterotrophs) get their matter and energy from their food (other organisms or bits of organisms that they "eat").
- Animals/heterotrophs need to break down the macromolecules in their food to monomers that can be taken up by their cells. This is the process of digestion.
- All organisms use the chemical potential energy in organic molecules as an energy source for cellular work. In the presence of oxygen, this work is most often cellular respiration.

Conceptual Barriers

Common Problems in Understanding

Many of the issues that students have understanding metabolism arise because they do not apply the laws of matter, energy, and conservation in their accounts of phenomena related to metabolism. Instead, students may see actors (plants, animals) who need enablers (Sun, food) to live or perform more specific functions. If the enablers are present, the actors will do what they're supposed to do, such as grow. Invisible substances (usually gases) may not be a part of their accounts. Even when they start to trace matter and energy, students often do so inconsistently. For example, they talk of food as being a source of energy that is "used up" by an organism (Parker, de los Santos, and Anderson 2015).

A number of commonly used phrases and words contribute to students' misunderstandings about metabolism.

- "I'm running out of energy" implies that energy is used up.
- "The food was turned into energy and used up" implies that matter can be turned into energy, which disappears.
- Plant "food" does not fit the scientific definition of food.
- Many careless statements exist in popular media; for example, "After all, not only does photosynthesis spin sunlight and water into the sugars we eat, it spawns as a happy waste product the oxygen we breathe" (Angier 2015). This account of photosynthesis from the *New York Times* does not include a carbon source.

Common Misconceptions

The misconceptions listed here do not compose an exhaustive list.

COMMON MISCONCEPTIONS ABOUT CELLULAR RESPIRATION

- Cellular respiration is the same as digestion.
- Cellular respiration is the process whereby an organism takes up oxygen and releases CO_2.
- During cellular respiration, food molecules are burned for energy and used up.
- Organisms lose weight by sweating or excreting matter or by using it up.
- Plants do not use cellular respiration.

COMMON MISCONCEPTIONS ABOUT PHOTOSYNTHESIS

- Photosynthesis is the plant equivalent of cellular respiration.
- Plants get their food from the soil.
- Gases cannot be a source of mass for plants.
- Plants can grow (add mass) in the dark using the dark reactions.

COMMON MISCONCEPTIONS ABOUT DIGESTION AND BIOSYNTHESIS

- As is cellular respiration, digestion is a process that yields energy.
- Digestion breaks down food into its component molecules, such as proteins or starches (not monomers).
- Cells can absorb food molecules.

- Cells absorb the molecules they need for growth and repair and use them unchanged.

Interdisciplinary Connections

A number of interdisciplinary connections can be made with the problems in this chapter. For example, connections can be made to language arts, geography, social studies, mathematics, and technology (see examples in Box 8.1).

Box 8.1. Sample Interdisciplinary Connections for Cellular Metabolism Problems

- **Language arts:** Write a verbal description of the changes in matter and conversion of energy in production of heat or electricity for home use. Write a story tracing an atom of carbon from the air taken in by a plant through the CO_2 you exhale.
- **Geography:** Explore the types of fuels used in different countries using geographic information systems technology.
- **Social studies:** Examine the historical use of fuels and how it affected developments in technology and the economy.
- **Mathematics:** Compare the calorie output of various foods using data from food labels. Create a calorie balance sheet of energy input and use for a person's data. Use different graphing formats to represent use or intake of energy.
- **Technology:** Use infrared temperature sensors to measure the temperature of burning marshmallows or wood to collect data for investigation.

References

Angier, N. *New York Times*. 2015. Our Ever Green World. April 20.

NGSS Lead States. 2013. *Next Generation Science Standards: For states, by states*. Washington, DC: National Academies Press. *www.nextgenscience.org/next-generation-science-standards.*

Parker, J. M., E. X. de los Santos, and C. W. Anderson. 2015. Learning progressions and climate change. *American Biology Teacher* 77 (4): 232–238.

Cellular Metabolism Problem 1: Torching Marshmallows

Alignment With the *NGSS*

PERFORMANCE EXPECTATIONS	• *MS-LS1-7:* Develop a model to describe how food is rearranged through chemical reactions forming new molecules that support growth and/or release energy as this matter moves through an organism. • *HS-LS1-7:* Use a model to illustrate that cellular respiration is a chemical process whereby the bonds of food molecules and oxygen molecules are broken and the bonds in new compounds are formed resulting in a net transfer of energy.
SCIENCE AND ENGINEERING PRACTICES	• Developing and Using Models • Analyzing and Interpreting Data • Constructing Explanations and Designing Solutions • Obtaining, Evaluating, and Communicating Information

DISCIPLINARY CORE IDEAS	• *LS1.C-MS Organization for Matter and Energy Flow in Organisms:* Within individual organisms, food moves through a series of chemical reactions in which it is broken down and rearranged to form new molecules, to support growth, or to release energy.
	• *LS1.C-HS Organization for Matter and Energy Flow in Organisms:*
	  • The sugar molecules thus formed contain carbon, hydrogen, and oxygen: their hydrocarbon backbones are used to make amino acids and other carbon-based molecules that can be assembled into larger molecules (such as proteins or DNA), used for example to form new cells.
	  • As matter and energy flow through different organizational levels of living systems, chemical elements are recombined in different ways to form different products.
	  • As a result of these chemical reactions, energy is transferred from one system of interacting molecules to another. Cellular respiration is a chemical process in which the bonds of food molecules and oxygen molecules are broken and new compounds are formed that can transport energy to muscles. Cellular respiration also releases the energy needed to maintain body temperature despite ongoing energy transfer to the surrounding environment.

CROSSCUTTING CONCEPTS	• Patterns • Cause and Effect: Mechanism and Explanation • Energy and Matter: Flows, Cycles and Conservation

Keywords and Concepts

Cellular respiration

Problem Overview

A boy wonders how burning a log in a campfire compares to eating a marshmallow. Full-color versions of all the problem's images are available on the book's Extras page at *www.nsta.org/pbl-lifescience*.

CELLULAR METABOLISM PROBLEM 1

Page 1: The Story

Torching Marshmallows

It was finally dry enough for Abe, who was enjoying his first day of summer vacation, to have a little campfire and toast and roast his favorite foods—marshmallows and potatoes. His parents had a fire pit in the backyard (Figure 8.1), not too far from an oak tree that had grown at least 20 feet since he first remembered seeing it. One of the larger branches had fallen down over the winter, so he chopped it up and carried the pieces to the fire pit. "Boy, these are really heavy!" Abe thought as he carried them. "How does a tree gain this kind of weight?"

Figure 8.1. Roasting Marshmallows

Being a good Boy Scout, he quickly had a fire going. Also being a good Scout, he wrapped his potato in tinfoil and threw it in the fire to eat later. Not always being the perfect Scout, he completely torched his first marshmallow—the small piece left was completely black! Abe began to think about what he had just observed. He had learned in school the basic principle that energy and matter are conserved but can change. He wondered: What had happened to the "missing" marshmallow? As the marshmallow burned, he noticed some forms of energy released, including heat, light, and sound. "Where did that energy 'come from'?" he asked himself as he prepared another marshmallow. This one came out perfectly. He popped it in his mouth. "Hmmm … I wonder what happens to the matter and energy in the marshmallow in my body? It can't be the same as completely torching it, or is it?"

Your Challenge: *What happens to the matter and energy in a marshmallow when you eat it? In what ways are these matter and energy changes similar to and different from burning the marshmallow?*

CELLULAR METABOLISM PROBLEM 1

Page 2: More Information

Torching Marshmallows

Abe remembered a lot of terms from his science class that he thought might be relevant to his questions, but he couldn't remember what they meant. He did a little online research, and the information below is what he found. Your task is to help Abe connect these terms and concepts to his questions about burning and eating marshmallows.

- "The *law of conservation of energy* states that the total energy of an isolated system remains constant—it is said to be *conserved* over time. Energy can neither be created nor destroyed; rather, it transforms from one form to another." (*en.wikipedia.org/wiki/Conservation_of_energy*)
- In everyday situations, *matter is conserved* during chemical reactions. All of the atoms in the starting substances are present in the products.
- Combustion is "a chemical reaction between substances, usually including oxygen and usually accompanied by the generation of heat and light in the form of flame." (*www.britannica.com/EBchecked/topic/127367/combustion*)
- A marshmallow is "a sugar-based confection that, in its modern form, typically consists of sugar, water and gelatin whipped to a spongy consistency, molded into small cylindrical pieces." (*https://en.wikipedia.org/wiki/Marshmallow*) See Figure 8.2 for a molecular model.
- Food is a material "used in the body of an organism to sustain growth, repair, and vital processes and to furnish energy." (*www.merriam-webster.com/dictionary/food*)
- Digestion is "the process of making food absorbable by dissolving it and breaking it down into simpler chemical compounds that occurs in the living body chiefly through the action of enzymes." (*www.merriam-webster.com/dictionary/digestion*). Another definition of *digestion* is "the breakdown of large insoluble food molecules into small water-soluble food molecules so that they can be absorbed into the watery blood plasma. In certain organisms [such as humans], these smaller substances are absorbed through the small intestine into the blood stream." (*http://en.wikipedia.org/wiki/Digestion*)
- Cells absorb the molecules they need from the blood. They excrete molecules that they don't need into the blood.

Abe still remembered the meaning of one term because it had puzzled him at first. His biology teacher had talked about *fixed carbon*, and Abe sat through a whole class period wondering what was wrong with the carbon that it needed fixing. He asked his mom when he got home, and after laughing a bit, she explained that fixed carbon was the same as organic carbon—carbon that was bonded to other carbon atoms and/or hydrogen. His mom said that he could think of fixed carbon compounds as being "fixed" because they have more potential energy than inorganic carbon compounds such as carbon dioxide (CO_2).

Figure 8.2. Molecular Model of the Sugar Glucose ($C_6H_{12}O_6$)

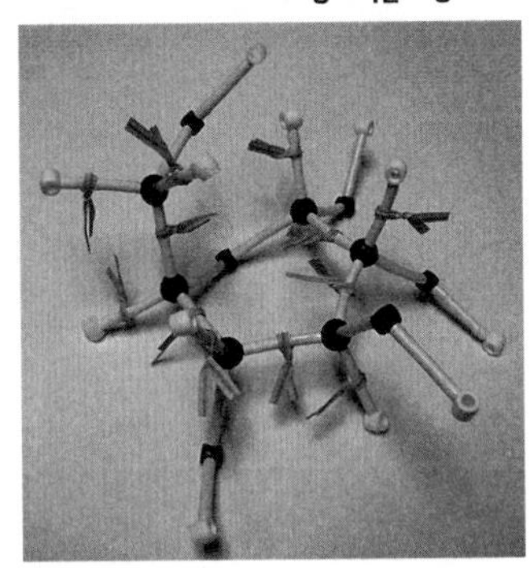

Note: A full-color version of this figure is available on the book's Extras page at *www.nsta.org/pbl-lifescience*.

Your Challenge: *What happens to the matter and energy in a marshmallow when you eat it? In what ways are these matter and energy changes similar to and different from burning the marshmallow?*

CELLULAR METABOLISM PROBLEM 1

Page 3: Investigations

Torching Marshmallows

Burning Marshmallows Investigation

MATERIALS

- Dry marshmallows
- Digital balance that measures to 0.1 g (to tell if there is a net gain or loss of atoms in the reaction mix)
- Plastic petri dish
- Large paper clip
- Masking or duct tape
- Matches or lighter
- Bromthymol blue solution (BTB) (blue under standard conditions; green or yellow in the presence of excess CO_2)
- Glass or foil-lined plastic container to cover reaction
- Indirectly vented chemical splash goggles, aprons, and nonlatex gloves for each student

PROCEDURE

(See Figure 8.3 for a photo of the setup.)

1. Bend a large paper clip so that the inside loop sticks out from the outside loop. Bend the end of the smaller inside loop so that it sticks out over the larger outside loop.
2. Carefully push a *dry* marshmallow onto the short end of the paper clip. Carefully tape the large loop of the paper clip to the lid of the petri dish.
3. Turn the digital balance on and place the petri dish with the paper clip and marshmallow onto the balance. Record the mass.
4. Place the bottom of another petri dish with about 25 ml of blue BTB in it next

to the digital balance. Record the color of the BTB.

5. Test to make sure you can cover the balance and the dish of BTB with the large container. Have your teacher light the marshmallow on fire and cover the setup with the container.

6. Observe the reaction until the flame goes out. Then, let the reaction sit undisturbed for an additional 15–20 minutes.

Figure 8.3. Burning Marshmallows Investigation

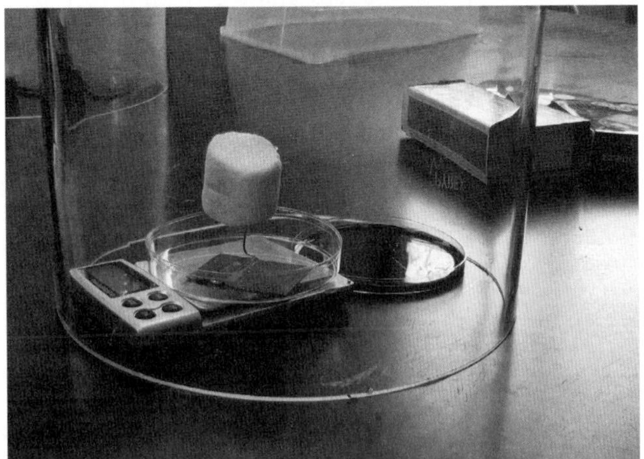

7. Remove the container. Record the color of the BTB and the mass of the marshmallow reaction setup. If the digital balance has turned off, gently take the petri dish off of it (being careful to not disturb the charred marshmallow), turn the balance on, put the petri dish with the marshmallow back on it, and record its mass.

SAFETY PRECAUTIONS

- Remind students not to eat any food brought into the lab or used in the lab activity.
- Keep all flammable and combustible materials away from active flames in the lab.
- Tell students to be careful when working with active flames so that they do not get burned.
- Make sure long hair is tied back.
- Review the Safety Data Sheet for BTB and share safety precautions with students.
- Appropriately dispose of all products from the activity.
- Have students wash their hands with soap and water after completing the activity.

Mealworm Investigation

This investigation is from Carbon TIME and is available at *http://carbontime.bscs.org*.

MATERIALS

- Plastic tub, small terrarium, or other container for mealworms
- Slices of fresh potato
- 100–150 mealworms
- Digital balance that measures to 0.1 g (to tell if there is a net gain or loss of atoms in the reaction mix)
- BTB (blue under standard conditions; green or yellow in the presence of excess CO_2)

PROCEDURE

1. Get a small container to hold your mealworms during the investigation. Make sure the container is deep enough that the mealworms cannot crawl out, and make sure the container has holes in the lid for ventilation so your mealworms have air.

2. If your mealworms are already in their meal bedding and container, you will need to separate the mealworms from the bedding. Using the end of a pencil, separate all the worms from the meal. If your worms come already separated from bedding, skip this step.

3. Place an empty small container onto the digital balance and "zero" out the scale. Then gently pour about 15 g of mealworms into this container. Record the mass of the mealworms.

4. Measure the mass of a slice of potato and record its mass.

5. Place the potato slice in the container with the mealworms (see Figure 8.4).

6. Place a petri dish with about 25 ml of blue BTB in it near the container with the mealworms. Record the color of the BTB. Cover this setup with a large container.

Figure 8.4. Mealworm Investigation

7. After 24 hours, measure the mass of the potato and the mass of the mealworms. Use the same zeroing procedures as above. Record both masses. Observe the color of the BTB.

CELLULAR METABOLISM PROBLEM 1

Teacher Guide

Torching Marshmallows

Problem Context

This problem focuses on combustion and metabolism of marshmallows. Marshmallows are a food that is mostly sugar and that students may have experience burning. Alternative contexts include burning wood, oil, a cracker, or a nut.

Model response: Burning (combustion) and metabolizing (cellular respiration) sugar are the same basic oxidation reaction:

$$C_6H_{12}O_6 + 6O_2 \rightarrow 6CO_2 + 6H_2O$$

This equation explains how the matter changes. Oxygen from the air combines with the sugar, producing CO_2 and water. The burnt marshmallow lost mass because the CO_2 produced as it burned escaped into the air. Some of the CO_2 turned the BTB yellow. The water produced was visible as condensation on the inside of the container after the marshmallow burned. The mealworms lost mass because they were respiring and the CO_2 produced escaped into the air. Some of the CO_2 turned the BTB yellow.

To account for the energy, we need to look at the bonds in the starting and ending substances. In the starting substances, there are many high-energy bonds (C–C, C–H, and O–O). It takes only a little energy to break them. The new bonds formed in the products (C–O and H–O) are more stable, and much energy is released when they form. Thus, the net effect is release of energy. In the case of combustion, the energy is released as heat and light. In the case of cellular respiration, some of the energy is captured as chemical potential energy in ATP, which is used to do the work of the cell. Eventually, ATP ends up as heat.

Activity Guide

The definitions of *combustion, food,* and *digestion* on Page 2 of the problem are meant to be a response to words that students are likely to bring up when thinking about Page 1. If they do not bring up these topics, the definitions can be used to broaden their thinking. Students are not expected to think of the conservation laws, chemical formulas, or bonds on their own. These pieces of information are meant to guide students into thinking chemically about cellular respiration.

The investigations on Page 3 can be used in more or less structured ways (SEP 3: Planning and Carrying Out Investigations). Students can use the information in the list of materials to develop a way of investigating combustion or respiration. However, it is more realistic to give students the protocol for the combustion investigation and to discuss the reasoning behind it. Careful observation of the reaction will help lead students to the equation. For example, condensation will form on the sides of the container—evidence that water is produced (see Figure 8.5). After that experience, students might be able to suggest procedures for the mealworm investigation. Alternatively, you can give them the protocol.

Figure 8.5. Burning Marshmallow and Condensation

To help students visualize the matter and energy, they can model the reaction using models like the one shown in Figure 8.2 (p. 193) (SEP 2: Developing and Using Models). An easy way to do this is to have each group model the starting substances (such as glucose and oxygen), placing twist ties on the high-energy bonds (C–C, C–H, O–O). Then two groups can combine and rearrange one of their sets to form the products simulating the reaction. They will have twist ties left over, representing the released energy. Students can count the atoms of each type in the reactants and products to confirm that matter has been conserved.

CELLULAR METABOLISM PROBLEM 1

Assessment

Torching Marshmallows

Transfer Task

"Make hay while the sun shines" is an old saying that really does apply to farming. When farmers cut a hay field and make hay bales (Figure 8.6), they want the grass to dry as quickly as possible. When the grass is dry, it rots (decomposes) much more slowly, and the farmers can store the dried grass or hay to feed animals in the winter. If the grass doesn't dry fast enough, it decomposes. Decomposition happens because microbes on the grass use it as a food source for cellular respiration.

Figure 8.6 Hay Bale

Hay that is not properly dried (cured) can get very hot and even catch on fire. Occasionally barns full of improperly cured hay burn down. (It may be helpful to know that grass is mostly cellulose, a polymer of sugar.)

- Explain why the uncured grass gets so hot. Your answer should explain at the molecular level what is happening.
- In addition to temperature, what can farmers monitor to tell if their hay has cured properly and is safe to store? Explain your answer.

Model response: The grass gets so hot because the microbes on it are doing cellular respiration. They use the sugar in the cellulose and combine it with oxygen. During this process, the microbes use a little energy to break the C–C, C–H, and O–O bonds in the sugar and oxygen. A lot more energy is released when stable C–O and H–O bonds in CO_2 and water are formed. The microbes convert some of this energy into chemical potential energy for their own use. When they use the chemical potential energy, it is eventually converted to heat [CC 3: Scale, Proportion, and Quantity; CC 5: Energy and Matter: Flows, Cycles, and Conservation].

In addition to temperature, farmers could monitor the amount of CO_2 produced by the grass. If the grass is producing a lot of CO_2, it is not dry enough to stop the microbes from doing cellular respiration [SEP 3: Planning and Carrying Out Investigations].

General Question

A pancreas cell needs energy and a carbon source so that it can synthesize insulin. A bone marrow cell (Figure 8.7) needs energy and a carbon source so that it can synthesize hemoglobin. Explain how cells get the energy and carbon sources they need to do their work.

Model response: Cells take up food molecules from the blood that goes by them. They use some of these molecules with appropriate combinations of atoms to make the molecules they need. They get the energy they need by doing cellular respiration. During this process, a little energy is used to break the C–C, C–H, and O–O bonds in the sugar and oxygen. A lot of energy is released when new stable C–O and H–O bonds in CO_2 and water are formed. Cells convert some of this to chemical potential energy for their own use. The rest is lost as heat [CC 5: Energy and Matter: Flows, Cycles, and Conservation; SEP 6: Constructing Explanations and Designing Solutions].

Figure 8.7. Bone Marrow Cells

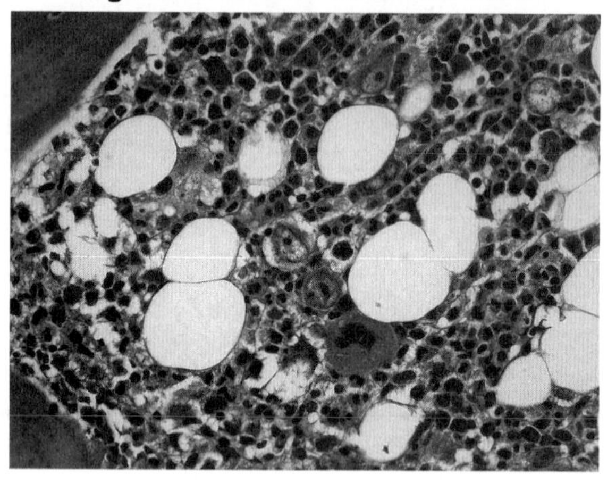

Application Questions

APPLICATION QUESTION 1

You have a friend who switched to a healthier diet with less fat and sugar and more fruit and vegetables (fiber, vitamins). The healthier diet also had fewer calories and she lost 10 pounds, which pleased her doctor. What happened to the mass/weight that she lost? It may be helpful to know that body fat is mostly molecules such as stearic acid, $C_{18}H_{36}O_2$.

Model response: Because my friend was eating fewer calories, her body started using her fat as an energy source. After her cells used all of the molecules from her food in cellular respiration, they used fat molecules if they still needed energy. Cellular respiration produces CO_2 and water. She breathed out the CO_2. The water was absorbed by her body [CC 5: Energy and Matter: Flows, Cycles, and Conservation; SEP 6: Constructing Explanations and Designing Solutions].

Note to teacher: Eventually, the water was also lost through exhalation or removed by her kidneys. The actual fate of the water goes beyond the scope of what students studying cellular respiration learn and does not need to be part of their answer.

APPLICATION QUESTION 2

A compost pile that sits for six months loses a lot of mass and produces heat. Explain what is happening at the molecular level.

> *Model response:* Fungi and microbes in the compost pile use the molecules in the decaying material of the compost pile as a carbon source and for the energy they need to build their bodies. If oxygen is available, the fungi and microbes are likely to undergo cellular respiration to transform the chemical potential energy in the decaying material into chemical energy they can use. Cellular respiration produces CO_2 and water, which can escape the compost pile, causing it to lose mass. The chemical potential energy in the decaying material is eventually transformed into heat, which warms the compost pile [CC 5: Energy and Matter: Flows, Cycles, and Conservation; SEP 6: Constructing Explanations and Designing Solutions].

Common Beliefs

Indicate whether the statements are true (T) or false (F), and explain why you think so (*model responses shown in italics*).

- During cellular respiration, sugar is converted into energy and used up. *(F) During cellular respiration, sugar is converted to CO_2 and water. The organism expels/breathes out the CO_2 (and much of the water).*
- During cellular respiration, the atoms from sugar end up in ATP. *(F) During cellular respiration, the atoms from sugar end up in CO_2 and water. ATP is made from adenosine diphosphate and phosphate. It is the chemical potential energy, not the atoms, that is transferred from the sugar to the ATP.*

Cellular Metabolism Problem 2: Mysterious Mass

Alignment With the *NGSS*

PERFORMANCE EXPECTATIONS	• *MS-LS1-6:* Construct a scientific explanation based on evidence for the role of photosynthesis in the cycling of matter and flow of energy into and out of organisms. • *HS-LS1-5:* Use a model to illustrate how photosynthesis transforms light energy into stored chemical energy.
SCIENCE AND ENGINEERING PRACTICES	• Developing and Using Models • Analyzing and Interpreting Data • Constructing Explanations and Designing Solutions • Obtaining, Evaluating, and Communicating Information

DISCIPLINARY CORE IDEAS	• *LS1.C-MS Organization for Matter and Energy Flow in Organisms:* Within individual organisms, food moves through a series of chemical reactions in which it is broken down and rearranged to form new molecules, to support growth, or to release energy.
	• *LS1.C-HS Organization for Matter and Energy Flow in Organisms:*
	  • The sugar molecules thus formed contain carbon, hydrogen, and oxygen: their hydrocarbon backbones are used to make amino acids and other carbon-based molecules that can be assembled into larger molecules (such as proteins or DNA), used for example to form new cells.
	  • As matter and energy flow through different organizational levels of living systems, chemical elements are recombined in different ways to form different products.
	  • As a result of these chemical reactions, energy is transferred from one system of interacting molecules to another. Cellular respiration is a chemical process in which the bonds of food molecules and oxygen molecules are broken and new compounds are formed that can transport energy to muscles. Cellular respiration also releases the energy needed to maintain body temperature despite ongoing energy transfer to the surrounding environment.

CROSSCUTTING CONCEPTS	• Patterns • Cause and Effect: Mechanism and Explanation • Energy and Matter: Flows, Cycles, and Conservation

Keyword and Concept

Photosynthesis

Problem Overview

A camper thinks about how the tree that produced the firewood he is burning created the material that makes up the tree. Full-color versions of all the problem's images are available on the book's Extras page at *www.nsta.org/pbl-lifescience*.

CELLULAR METABOLISM PROBLEM 2

Page 1: The Story

Mysterious Mass

One night while camping, Abe's fire went out. He had baked a potato in the fire for dinner. He removed the potato and started to clear out the ashes. He lifted a shovel full. "Wow—this stuff is really light!" He was immediately reminded of how heavy the wood was when he filled the fire pit. "I know that wood loses mass when it is burned, just like when our bodies burn food for energy. But, I wonder, how did the wood get so heavy in the first place?"

Your Challenge: *In what ways are the matter and energy changes in burning wood similar to and different from those in burning food for energy? What are the matter and energy changes that occur when a plant grows?*

CELLULAR METABOLISM PROBLEM 2

Page 2: More Information

Mysterious Mass

Abe didn't really know what wood is made of, so he did some online research and found this definition: "*Wood* is a porous and fibrous structural tissue found in the stems and roots of trees and other woody plants. It has been used for thousands of years for both fuel and as a construction material. It is an organic material, a natural composite of cellulose" and other molecules (*http://en.wikipedia.org/wiki/Wood*). There are several definitions of *cellulose*, but a useful one for the purpose of this problem is as follows: "a rigid, colorless, unbranched, insoluble, long-chain polysaccharide, consisting of 3,000 to 5,000 glucose [sugar] residues and forming the structure of most plant structures and of plant cells" (*http://medical-dictionary.thefreedictionary.com/cellulose*). Figure 8.8 shows a molecular model of glucose.

Figure 8.8. Molecular Model of the Sugar Glucose ($C_6H_{12}O_6$)

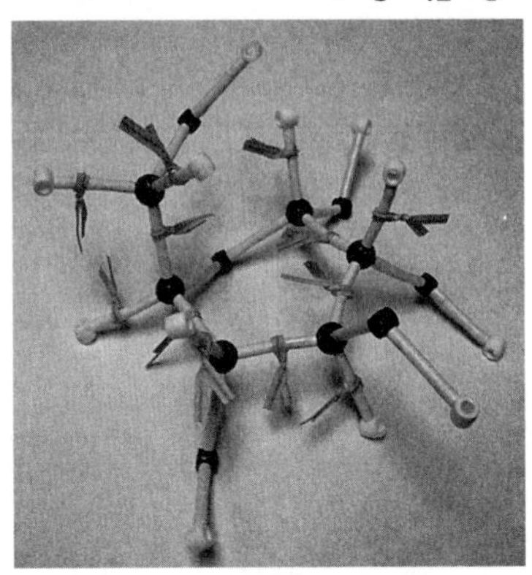

Note: Large black beads = C atoms; smaller black beads = O atoms; white beads = H atoms; white sticks = chemical bonds; yellow twist ties = high-energy bonds. A full-color version of this figure is available on the book's Extras page at *www.nsta.org/pbl-lifescience*.

Your Challenge: *In what ways are the matter and energy changes in burning wood similar to and different from those in burning food for energy? What are the matter and energy changes that occur when a plant grows?*

CELLULAR METABOLISM PROBLEM 2

Page 3: Resources and Investigations

Mysterious Mass

Resources

1. Carbon TIME plant investigation video of plants producing oxygen in the light and CO_2 in the dark: *http://carbontime.bscs.org*. Register and go to Activity 2.2 in the Plants unit.

2. Time lapse of acorn growing: *www.youtube.com/watch?v=4BtKAkP5xOk*.

Investigation

MATERIALS

- A thick dry twig set in a small beaker
- Digital balance that measures to 0.1 g (to tell if there is a net gain or loss of atoms in the reaction mix)
- Matches or lighter
- Plastic petri dish
- BTB (blue under standard conditions; green or yellow in the presence of excess CO_2)
- Glass or foil-lined plastic container to cover reaction
- Indirectly vented chemical splash goggles, aprons, and nonlatex gloves for each student

PROCEDURE

1. Turn the digital balance on and place the beaker with the twig onto the balance. Record the mass.

2. Place the bottom of the petri dish with blue BTB in it next to the digital balance. Record the color of the BTB.

3. Test to make sure you can cover the balance, the beaker, and the dish of BTB with

the large container. Have your teacher light the twig on fire and cover the setup with the container.

4. Observe the reaction until the flame goes out. Let the reaction sit undisturbed for an additional 15–20 minutes.

5. Remove the container. Record the color of the BTB and the mass of the twig reaction setup. If the digital balance has turned off, gently take the petri dish off of it (being careful to not disturb the charred twig), turn the balance on, put the petri dish with the twig back on it, and record its mass.

SAFETY PRECAUTIONS

- Keep all flammable and combustible materials away from active flames in the lab.
- Tell students to be careful when working with active flames so that they do not get burned.
- Make sure long hair is tied back.
- Review the Safety Data Sheet for BTB, and share safety precautions with students.
- Appropriately dispose of all products from the activity.
- Have students wash their hands with soap and water after completing the activity.

CELLULAR METABOLISM PROBLEM 2

Teacher Guide

Mysterious Mass

Problem Context

As with Torching Marshmallows, the context for this problem—burning wood—was chosen as something familiar to students. However, other examples of combustion of organic materials can be used, such as burning paper or candles.

> *Model response:* The matter and energy changes in burning wood and burning marshmallows are essentially the same. In both cases, the substance burning is sugar. The sugar combines with oxygen and produces CO_2 and water. The evidence for this is that the wood and the marshmallow both lose mass when they are burned and BTB turns green or yellow, indicating that CO_2 is given off. The chemical potential energy in the sugar and oxygen is converted to heat and light.
>
> The matter and energy changes in growing plants are the reverse of those in cellular respiration. In the light, plants convert CO_2 from the air and water into sugar and oxygen. The energy in the sunlight is converted to chemical potential energy in the C–C, C–H, and O–O bonds. The sugar is used for building the plant and meeting its energy needs. The evidence for this is that plants in the light gain weight and that yellow BTB turns green or blue, indicating that CO_2 is being taken up.

Activity Guide

As with Torching Marshmallows, the investigations on Page 3 can be used in more or less structured ways. Students can use the information in the list of materials to develop a way of investigating combustion of wood. If they had experience doing the investigations in the Torching Marshmallows problem, they may be able to figure out a similar protocol for this reaction. Alternatively, if you would like to shorten the time needed, you can give them the protocol.

To help students visualize the matter and energy changes in respiration and photosynthesis and compare them, they can model the reaction using models such as the one shown on Page 2 (SEP 2: Developing and Using Models). The combustion reaction is exactly the same as what they modeled in Torching Marshmallow (see the Activity Guide, p. 198, for that problem). To help students work out the reaction for photosynthesis, it may be helpful to have them leave out their models of respiration for reference. Then, have them model the products of photosynthesis (sugar and oxygen) and compare these to the respiration reaction. They should see that the products of photosynthesis are the same as the reactants for respiration. That one reaction is the reverse of the other is confirmed by their BTB results.

CELLULAR METABOLISM PROBLEM 2

Assessment

Mysterious Mass

General Questions

GENERAL QUESTION 1

An investigator placed 1.5 g of radish seeds (dry) in a petri dish and provided the seeds with only light and water. After one week, the material in the dish was dried and weighed. The material weighed 3.28 g. Where did the mass come from?

> *Model response:* The seeds used the light and water, along with CO_2 from the air, in the process of photosynthesis to produce sugars. They used these sugars as the basis for making new plant material and as a source of energy [CC 5: Energy and Matter: Flows, Cycles, and Conservation].

GENERAL QUESTION 2

What is wrong with this statement taken from a 2015 science news article by Natalie Angier in the *New York Times?* "After all, not only does photosynthesis spin sunlight and water into the sugars we eat, it spawns as a happy waste product the oxygen we breathe."

> *Model response:* This description of photosynthesis makes it sounds as if sugars (which contain carbon) can be made without a carbon source. This description leaves out CO_2. It violates the conservation of mass [CC 5: Energy and Matter: Flows, Cycles, and Conservation].

Application Question

Figure 8.9 shows the level of CO_2 in the atmosphere as measured at the observatory at the top of Mauna Loa in Hawaii. The solid line tracks the trend over time, whereas the wavy gray line shows the monthly average. Most places in the Northern Hemisphere show a similar trend, with the highest levels coming in May and the lowest in October. Why is this?

> *Model response:* During May and October, plants are active and growing. They undergo photosynthesis, which uses CO_2 from the air [SEP 4: Analyzing and Interpreting Data].

Common Beliefs

Indicate whether the statements are true (T) or false (F), and explain why you think so (*model responses shown in italics*).

1. Plants get their mass from substances in the soil and water. *(F) Plants take only a small amount of matter from the soil, mostly in the form of nitrogen and phosphate. Most of their mass comes from CO_2 in the air and water.*

2. Plants make oxygen for us. *(F) Oxygen is simply a by-product of photosynthesis as sugar is made. Plants must undergo photosynthesis to have energy and building blocks for growth. They also undergo cellular respiration and use oxygen.*

Figure 8.9. Level of Carbon Dioxide Measured at Mauna Loa Observatory in Hawaii

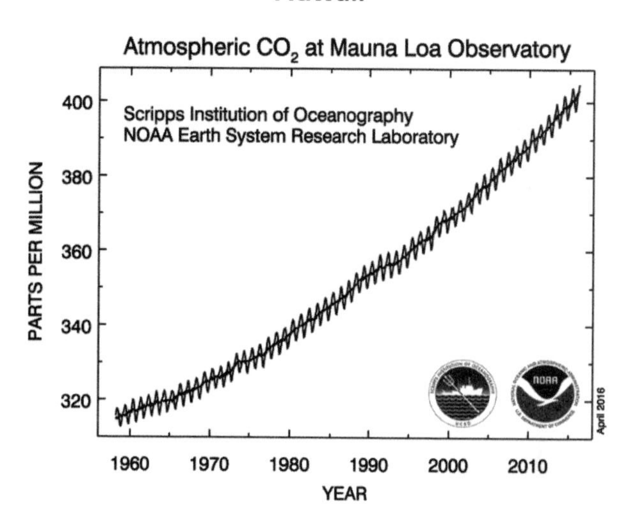

Cellular Metabolism Problem 3: Why We Are Not What We Eat

Alignment With the *NGSS*

PERFORMANCE EXPECTATIONS	• *MS-LS1-7:* Develop a model to describe how food is rearranged through chemical reactions forming new molecules that support growth and/or release energy as this matter moves through an organism. • *HS-LS1-6:* Construct and revise an explanation based on evidence for how carbon, hydrogen, and oxygen from sugar molecules may combine with other elements to form amino acids and/or other large carbon-based molecules. • *HS-LS1-7:* Use a model to illustrate that cellular respiration is a chemical process whereby the bonds of food molecules and oxygen molecules are broken and the bonds in new compounds are formed resulting in a net transfer of energy.
SCIENCE AND ENGINEERING PRACTICES	• Developing and Using Models • Analyzing and Interpreting Data • Constructing Explanations and Designing Solutions • Obtaining, Evaluating, and Communicating Information

DISCIPLINARY CORE IDEAS	• *LS1.C-MS: Organization for Matter and Energy Flow in Organisms:* Within individual organisms, food moves through a series of chemical reactions in which it is broken down and rearranged to form new molecules, to support growth, or to release energy.
	• *LS1.C-HS: Organization for Matter and Energy Flow in Organisms:*
	  • The sugar molecules thus formed contain carbon, hydrogen, and oxygen: their hydrocarbon backbones are used to make amino acids and other carbon-based molecules that can be assembled into larger molecules (such as proteins or DNA), used for example to form new cells.
	  • As matter and energy flow through different organizational levels of living systems, chemical elements are recombined in different ways to form different products.
	  • As a result of these chemical reactions, energy is transferred from one system of interacting molecules to another. Cellular respiration is a chemical process in which the bonds of food molecules and oxygen molecules are broken and new compounds are formed that can transport energy to muscles. Cellular respiration also releases the energy needed to maintain body temperature despite ongoing energy transfer to the surrounding environment.

CROSSCUTTING CONCEPTS	• Patterns • Cause and Effect: Mechanism and Explanation • Energy and Matter: Flows, Cycles, and Conservation

Keywords and Concepts

Digestion, biosynthesis

Problem Context

A boy wonders how our body changes the materials of things we eat into our own tissue. Full-color versions of all the problem's images are available on the book's Extras page at *www.nsta.org/pbl-lifescience.*

CELLULAR METABOLISM PROBLEM 3

Page 1: The Story

Why We Are Not What We Eat

Abe cooked a potato on the grill in the backyard and took it into the house to put butter, salt, and pepper on it (Figure 8.10). As he passed by the garden plot, he recalled recently helping his mom plant potatoes. They had cut up potatoes, much like the one he had just roasted, making sure that each piece had an "eye" which would become a new potato plant. That was only a few weeks ago and some of those eyes had sprouted, and one sprout even had leaves on it! It was clear that a whole new plant was growing without sunshine. Abe thought, "It looks like a potato can be food for me and food for a plant. Does that mean that *food* is the same for both people and plants? The potato grows a full potato plant, but I don't turn into a potato or a potato plant when I eat potatoes. Do plants and animals use the molecules in food in the same or different ways? I'll have to figure this out. For now, I'm going to enjoy my fancy potato!"

Figure 8.10. Baked Potato

Your Challenge: *How do animals and plants use food? Why do we not become what we eat?*

CELLULAR METABOLISM PROBLEM 3

Page 2: More Information

Why We Are Not What We Eat

Abe followed up his wondering with some quick online research. He learned the following information:

- *Food* is a material "used in the body of an organism to sustain growth, repair, and vital processes and to furnish energy" (*www.merriam-webster.com/dictionary/food*).
- *Digestion* is "the process of making food absorbable by dissolving it and breaking it down into simpler chemical compounds that occurs in the living body chiefly through the action of enzymes" (*www.merriam-webster.com/dictionary/digestion*). Another definition of *digestion* is "the breakdown of large insoluble food molecules into small water-soluble food molecules so that they can be absorbed into the watery blood plasma. In certain organisms [such as humans], these smaller substances are absorbed through the small intestine into the blood stream" (*http://en.wikipedia.org/wiki/Digestion*).
- Cells absorb the molecules they need from the blood. They excrete molecules that they don't need into the blood.

Abe also found another useful resource—a time-lapse video of a sprouting potato: *www.youtube.com/watch?v=EGx7rW_elok.*

Your Challenge: *How do animals and plants use food? Why do we not become what we eat?*

CELLULAR METABOLISM PROBLEM 3

Page 3: Resources

Why We Are Not What We Eat

1. Carbon TIME plant investigation video of plants producing oxygen in the light and CO_2 in the dark: *http://carbontime.bscs.org*. Register and go to Activity 2.2 in the Plants unit. This video, which is also used in the Mysterious Mass problem, shows what happens to mass and CO_2 when plants are in the light (doing more photosynthesis than cellular respiration) and when plants are in the dark (doing only cellular respiration). Plants also do cellular respiration when they are in the light, but because the rate of photosynthesis is greater than that of respiration, the net effect is intake of CO_2 in the light.

2. Biosynthesis pathways chart: Many of these charts are available online. One example is *www.sigmaaldrich.com/technical-documents/articles/biology/interactive-metabolic-pathways-map.html.* The point of the biosynthesis pathways chart is not for students to look at individual reactions but for them to get the idea that cells have multiple pathways for making the molecules they need from what's available.

3. See the Potato Respiration investigation in the Activity Guide (p. 218) for an inquiry activity to help students experiment with biosynthesis and metabolism.

CELLULAR METABOLISM PROBLEM 3

Teacher Guide

Why We Are Not What We Eat

Problem Context

Students often wonder if the saying "you are what you eat" makes sense. If they eat vegetables, they are not becoming vegetables. Students can observe a difference in structure between their body and a vegetable. Often, this observation leads to questions and doubts about what happens to the foods they eat and how their body can use food to grow. In this problem, Abe wonders about this, and he investigates how his body can break down the molecules that make up his potato and turn it into the molecules that make up his own body. This problem can be modified to use any vegetable or fruit familiar to students.

Model response: Both animals and plants use food for growth, repair, and other cellular processes and as an energy source. "Food" molecules contain carbon and hydrogen and usually oxygen. Animals get food molecules from other organisms (or bits of organisms) that they eat. Fungi and many microbes also get their food molecules from other organisms that they ingest in various ways. Plants make their own food. When plant or animal cells, as well as those of microbes and fungi, use food molecules for energy, they use a small amount of energy to break the unstable, high-energy C–C and C–H bonds in the food and the O–O bonds in oxygen. They form stable, low-energy C–O and H–O bonds in CO_2 and water and use the residual energy to do cellular work. [See Cellular Metabolism Problem 2 for information about how cells convert chemical bonds into usable energy.]

The evidence that, like animals and fungi, plants do cellular respiration is from the Carbon TIME video. In the dark when plants can't do photosynthesis, they still do respiration. This is evident in their mass loss and CO_2 production. Students can confirm that potatoes undergo cellular respiration by doing the potato investigation.

Cells don't absorb whole molecules from their food and use them unchanged. Much of our food is polymers. During digestion, the molecules are broken down into smaller monomers. The monomers are absorbed into the bloodstream. Cells absorb the monomer molecules they need to do cellular processes or to "burn" for energy. If the monomers are not exactly what they need, cells can build them into other molecules.

As shown by the biosynthesis pathways chart (see Resources, p. 216), cells can do many, many reactions turning molecules into other molecules.

Activity Guide

Students can test to see if the potato is undergoing cellular respiration by doing an investigation with BTB and scales in the same way as the Mealworm Investigation (pp. 195–196) in the Torching Marshmallows problem.

Potato Respiration

MATERIALS

- Large plastic tub or container
- Lid or cover for the container
- Thin slices of potato
- Digital balance that measures to 0.1 g
- Plastic petri dish
- BTB (blue under standard conditions; green or yellow in the presence of excess CO_2)

PROCEDURE

1. Place multiple thin slices of potato standing up in a container (maximize contact with air).

2. Measure the mass of the potatoes and container to the nearest 0.1 g. Record the initial mass.

3. Place a petri dish with a small amount of the BTB solution in the container with the potatoes. The potatoes should not be touching the BTB.

4. Cover and let sit overnight.

5. The next day, record the color of the BTB solution. Explain what chemical might have caused the change in color and why that chemical was present.

6. Measure the final mass of the potato slices and container. If the mass changed, explain what might have caused the change in mass.

CELLULAR METABOLISM PROBLEM 3

Assessment

Why We Are Not What We Eat

Transfer Task

Mushrooms are fungi, not plants (Figure 8.11). They cannot make their own food. Mushrooms are not animals either. They don't have digestive systems. People grow mushrooms on dead plant matter. Mushrooms and other fungi contribute to the decay of the dead material they grow on. How do mushrooms get the molecules they need to grow?

Figure 8.11. Mushrooms

Model response: The fact that mushrooms cause decay implies that they secrete something that breaks down the molecules in the dead material they grow on. Instead of having a digestive system, they digest their food externally before they take it in. They absorb the broken-down molecules and use them for energy and to make the molecules they need to grow [CC 5: Energy and Matter: Flows, Cycles, and Conservation; SEP 6: Constructing Explanations and Designing Solutions].

General Question

You eat a protein bar. Later you do your pinkie exercises, working your little fingers up and down 100 times. Describe the matter and energy changes that the protein that you ate undergoes from the time you eat it to when it is used for energy by muscle cells in your little fingers.

Model response: In your digestive system, the protein is broken down into smaller molecules (amino acids), and the smaller molecules are absorbed into your blood. Digestion doesn't involve significant amounts of energy use or production. Muscle cells in your little fingers absorb the molecules from the blood and use a series of reactions to turn them into molecules that they can use in cellular respiration. During cellular respiration, a small amount of energy is used to break the unstable high-energy bonds (C–C, C–H, and O–O). Stable, low-energy bonds in CO_2 and water are formed, releasing a lot of energy. The muscle cells use some of this energy to contract. The energy ends up as heat [CC 5: Energy and Matter: Flows, Cycles, and Conservation; SEP 6: Constructing Explanations and Designing Solutions].

Application Question

Explain why a vegan (who eats only plant material) and someone who eats a lot of meat have essentially the same molecules in their bodies.

Model response: The molecules that both people eat are broken down in their digestive systems into simpler, smaller molecules that are absorbed in their blood. Their cells take molecules from the blood and use a series of reactions to turn them into the molecules they need. None of the large plant or animal molecules becomes part of the person [CC 5: Energy and Matter: Flows, Cycles, and Conservation; SEP 6: Constructing Explanations and Designing Solutions].

Common Beliefs

Indicate whether the following statements are true (T) or false (F), and explain why you think so (*model responses shown in italics*).

1. Plants do photosynthesis during the day and cellular respiration at night. *(F) Plants do need sunlight to do photosynthesis. However, they undergo cellular respiration all of the time, because they need energy to do cellular work all of the time.*

2. Plants gain mass by absorbing matter from the soil. *(F) Plants absorb only trace nutrients from soil. The vast majority of the mass they gain as they grow comes from the* CO_2 *in the air they "fix" during photosynthesis to build glucose and starch, including the cellulose that makes up most of the mass of a plant.*

3. The proteins in meats and vegetables we eat are exactly the same as the proteins that make up our body. *(F) When we eat meats and vegetables, or any food, our body breaks down the proteins and other molecules into smaller units and then rebuilds our own cellular molecules using those smaller units. If we eat proteins, our cells break them down into amino acids and use them to rebuild human proteins. So although we are made of the same amino acids, we have different proteins.*

4. Decomposing material loses mass by losing CO_2 and water. *(T) A compost pile loses mass over time. Decomposition happens when bacteria, fungi, and other organisms break down dead things and wastes, and they burn the molecules for energy. Cellular respiration is the reaction of burning food for energy, and this reaction converts large organic molecules into CO_2 and H_2O. The CO_2 is given off as a gas, and the water is given off as water vapor. So as decomposition gives off these products of respiration, the compost pile loses mass.*

MODIFYING AND DESIGNING YOUR OWN PROBLEMS

Chapters 5–8 presented 16 problems directly addressing four different life science content strands. These problems are a good way to introduce you and your students to problem-based learning (PBL), and we hope the lessons are useful in your classroom. But these problems are by no means enough to cover every concept in your curriculum. There are more science ideas than we can fit in this book that would make excellent PBL topics.

So you may want to consider developing your own PBL lessons, by modifying the ones we have presented, revising a current lesson plan to fit the PBL structure, or maybe even creating your own PBL problem from scratch. In this chapter, we will share some strategies for writing PBL lessons and tips for creating problems that are rich, engaging, and ideal for addressing the standards you need to teach. We have created a series of steps to think about as you design your own PBL lessons. The tips we provide are based on our own experience and the advice of classroom teachers who have successfully implemented their own PBL problems. The process we recommend also follows the principles of backward design described by Wiggins and McTighe (2005).

Selecting a Topic for a Story

One of the features of a PBL lesson that distinguishes this format from some other approaches is the importance of the context in which the problem is situated. The story, consisting of Page 1 and Page 2, is designed to engage students in the concepts of the lesson, help them recognize the practical relevance of the concept, and give them a reason to want to pursue the challenge and find a resolution to the problem. If you elect to develop your own PBL lessons, one of the issues you will need to address is how to write a story that will encompass these important features of a good PBL problem. So where do you begin?

For many teachers, especially those who plan to use the principles of backward design, the first step is usually to identify learning targets, align the lesson with standards, and write objectives. Although these are critical steps, there is an important place for inspiration as a starting point in the planning as well. As we discuss how to identify the story, we'll start with the inspiration or creative idea and then move toward identifying standards and learning goals as part of the process as well. In actual practice, the two ideas—the creative inspiration and the methodical planning for standards—are closely tied. We encourage you to seek a balance between the two.

Finding Inspiration for an Authentic Context

For many of the problems included in this book, a real event inspired the idea for the story. Some were written to reflect real events or a local problem or phenomenon that could be used to spark a science lesson. In other cases, the authors built a fictional—but realistic—story based on their understanding of the science concept and its practical implications. In developing your own problems, you will probably find that the best ideas share this authentic character. A real or authentic problem needs to be believable, or even based on a real problem!

That inspiration can come from several different sources. Sometimes a newspaper article, an item on a television news broadcast, or a story you hear on the radio may be the spark that starts a new PBL problem, such as the story of a pest damaging a local crop. Or maybe a real problem in your community, such as a controversy about what design to select for building a new sewage treatment plant, might be the inspiration. If your class works with a scientist from a business or university in your area, discussions with the expert in the field may lead to questions that lend themselves to the PBL format.

One valuable source of inspiration can be the questions students ask or their interests in science. Your students, especially in younger grades, are curious about the world around them, and they may bring ideas to class that would be excellent PBL problems. When a student brings a caterpillar to class and asks what it is, or asks about last night's story on the news about a lunar eclipse, you may be able to build your lesson from these questions. One advantage to this approach is that you can be certain that at least some students already want to know the answer before you start!

There are strategies you can use to help elicit students' interests and questions in the classroom. One is to ask students in the first days of your class to write down some "I wonder … " statements. These are short written comments about science questions they would like to explore. You can sort through these responses and find patterns in their interests, or maybe you can find a few outstanding questions you could develop into a PBL lesson.

One kindergarten teacher in the PBL Project for Teachers even created a Science Questions bulletin board in the classroom. During her weekly science lessons, any questions students raised could be written on a note card and thumbtacked to the bulletin board. She would then use those questions to design investigations for later lessons.

In other situations, you may need to build the contextualized problem. If you've already identified a standard and connected the concept to a context, you might be able to write a fictional but realistic scenario that presents a problem you can relate to your PBL plan. The genetics problems about coat color in cats (see Chapter 7) are examples of this process.

You may also find it necessary to combine some of these sources. One group of teachers in the PBL Project for Teachers designed a problem about the increasing growth of poison ivy in the Midwest for seventh- and eighth-grade science classes (Boersma, Ballor, and Graeber 2008). The original idea came from an online article that mentioned that poison ivy is growing faster and larger because of increasing levels of carbon dioxide (CO_2) in the atmosphere. The story addressed science concepts that fit perfectly in the eighth-grade standards about human influences on the environment, photosynthesis, and the carbon cycle.

To teach this lesson about poison ivy, Kylie (the same teacher described in Chapter 4, pp. 52–59) modified the problem for the students in her small rural community, where deer hunting is very popular. She decided to make the story about a student and her father going hunting and noticing that the poison ivy plants were larger than usual and found in more places.

Once she had identified her learning goals and an authentic problem, the rest of the process was easy. In the next sections, we will refer to Kylie's PBL lesson again to see how she created the elements she needed for her lesson.

Standards

Although you may begin your PBL planning with a story or context that inspires your idea, at some point you need to make sure that the lessons you develop address the standards that teachers are expected to cover. We suggest you follow the inspiration with some thought about your content standards.

Because you must account for students' learning of specific concepts, the content standards may be a good place to start planning. For some science concepts, you may have units in which you have strong inquiry-based activities and a plan that works very well for your students. A good way to start your new PBL plan might be to identify a unit that you feel needs to be revitalized. Or you may wish to choose a standard that you already teach well that you want to update or revise.

When you think you have a context or story that would fit in your curriculum, the next question should be whether the story actually addresses appropriate learning goals. For instance, if the story you select is based on a newspaper article about honeybees disappearing at the local apple orchard, you can search for standards that focus on symbiotic interactions between species. If there are standards that would include this concept, you have a story that fits your needs, and you can continue your lesson development. This topic also has the advantage of including other examples from different contexts, such as a tickbird and a rhinoceros, or a termite and the microbe in its gut that digests cellulose. For students to fully engage in the PBL problem, the story needs to be something they can relate to on a personal level, but teaching appropriate content is your primary responsibility.

Writing Assessments

In the backward design model (Wiggins and McTighe 2005), the next step after identifying a learning goal or objectives is to consider how to assess student learning. We suggest thinking about this before developing a story for the PBL problem you are working on, because knowing what student learning should look like will help you focus on the necessary questions and information for guiding students toward the goal.

Format of Questions and Assessments

Just as in any science unit, you can assess student learning in many ways. A test or a project is an effective way to assess certain types of learning, and these may be appropriate instruments for your PBL lesson. But for now, we will focus our attention on some of the types of questions and assessments described earlier in this book. We have found that certain types of assessment questions seem to be most effective in revealing students' thinking about the problems teachers present in a PBL lesson. Let's take a quick look at the type of assessments you might choose to write. For more detailed examples, please refer to Chapter 4.

GENERAL QUESTIONS

A general question is an open-ended response item that simply asks for a student's overall understanding of a concept. This type of question is very well suited for a pre/post assessment. An example that might work for Kylie's poison ivy problem is "How would changes in CO_2 levels affect the growth of plants?"

Although this type of question is good at identifying big gaps or fundamental misconceptions, no one question can tell a teacher everything about students' thinking. The general questions featured in Chapters 5–8 are helpful in assessing students' overall understanding and their ability to state connections between ideas. But most students' responses to these questions may cover several ideas at a very shallow level, especially if students perceive a time limit in answering the questions. They often try to touch on as many related ideas as possible, but they do not give in-depth explanations. The teacher should remember that general questions such as this example are not designed to probe or tease out the specific details that might be a barrier to understanding science concepts.

APPLICATION QUESTIONS

Although the general questions give you a glimpse into students' thinking, teachers still need a way to assess the depth of their understanding. In the PBL Project for Teachers, we explored some possible assessment strategies and eventually developed a structure for questions that ask learners to apply their ideas to authentic problems. The application question lets you ask an open-response question that elicits a slightly deeper understanding of a specific concept. The format for the question that gives the most information presents a situation to describe a phenomenon, asks for an explanation of how or why the events happen, and usually includes some key terms or ideas that should be included in the response. Here is an example for the poison ivy problem:

> NASA has investigated the use of plants as a source of food and oxygen in long-term space flight. They've found that plants grow faster in enclosed spaces if animals or bacteria are also growing in the same space. Explain why these plants grow faster when other kinds of organisms are present. Your response should include the materials plants need in order to carry out photo-synthesis.

You can see that the question gives some hints about the type of information relating to the scenario when it states what the response "should include." This may seem like it gives away the answer, but if the students do not have a deep understanding of the process of photosynthesis, their explanation will reveal which parts of the concept are unclear or confusing to them and which parts they simply have wrong.

Like the general questions, application questions can also be used in a pre/post assessment model. You may also want to include application questions if your goal is to have students understand the details of a concept, explain a concept in more than just general terms, or focus on a specific part of a more complex process. When used together, the general and application questions provide a coherent picture of what the learners know.

TRANSFER TASKS

Transfer tasks are questions that present the learner with the same concept as the PBL problem, but in a different context. The goal of this type of assessment is to identify the learner's ability to apply the new concept to other situations. This transfer of understanding is an important indicator of deep understanding (Bransford and Schwartz 1999).

The transfer tasks developed in the PBL Project for Teachers were worded as open-response questions that gave enough detail to help the reader understand the new scenario. An example can be seen in Transfer Task 2 of the Wogs and Wasps problem (pp. 81–82):

> Butterflies lay eggs that hatch into tiny caterpillars. The caterpillars eat (usually the plant on which they were hatched) and grow. When a caterpillar is big enough, it spins a chrysalis around itself. Inside the chrysalis, its body changes shape. When it comes out of the chrysalis, it is a butterfly with crumpled wings. The butterfly pumps fluid into the veins of its wings to straighten them out, and eventually it can fly away. As adults, many butterflies eat flower nectar.
>
> How are the life cycles of butterflies, wasps, and frogs alike and different?

This example is more explicit in helping the learner connect the new question with the original concept, but it is written for younger students. As with any assessment, you should base the level of your question on the learning goals and level of complexity you expect in the responses.

PROBLEM SOLUTION SUMMARIES

One way to determine if each group or individual student has learned the target concept is to ask for a written summary of the solution to the problem. The summaries elicit a coherent explanation of what solution the student feels will resolve the problem or challenge and why that explanation makes sense. As is the previous examples, this is an open-response question. See Chapter 4 (pp. 58–60) for an example of a solution summary.

Writing Model Responses

Once you've written assessment questions for your PBL problem, an important strategy is to write model responses. Think of this as the "answer key" to the assessment or as an ideal answer. The model response represents the type of complete and accurate answer you hope your students will be able to provide at the end of the lesson.

Writing this response early in the development process will help you focus your planning on tasks that help learners move toward successful achievement of the learning goals you have identified in the first part of your lesson plan. Some teachers will highlight key elements of the model response or mark those elements with a boldface font to help organize their evaluation of student responses. If you are using the model response to assign points for a grade, this can also help you identify the points you award for an answer.

Chapters 5–8 included many examples of model responses written for each of the problems. Use these as examples as you write your own assessments.

Writing the Story

Writing Page 1

With a story context in mind, the learning goals identified, and a plan for assessing learning, you can now write the story. In the model we adopted from partners in the field of medical education, the story consists of Page 1 and Page 2. Page 1 is the initial introduction, and Page 2 is presented to learners after analyzing Page 1 and generating lists of "known" information, questions, and hypotheses based on the story.

When you read examples of the stories from Chapters 5–8, they look very simple to write. They are short, and they include limited information. But in practice, the process of writing the story can prove more challenging than it looks!

One of the challenges, especially with Page 1, is determining how much information to leave out and how much extraneous information to include. Remember that the PBL story needs to be "ill-defined" in that information is missing. Some information is provided that will not help in finding a solution. Most teachers instinctively want to provide all the needed information. It is easy to give more than the learners truly need. But if you leave out too much information, students may not connect the context to the learning goals of the lesson. Finding a balance takes some careful thought and editing.

For Page 1, try to give just enough information to grab the attention of the learners. Page 1 of the story needs to describe the context and present some kind of question or challenge that needs to be answered. Let's take a look at an example from Page 1 of the Bogged Down problem in Chapter 6 (p. 133). In this problem, students should eventually focus on the idea that habitats change naturally over time through a process called ecological succession.

Page 1: The Story

Bogged Down

For the last 20 years, students have been visiting a bog located on the west side of the Rose Lake marsh in Clinton County, Michigan. Every year, more and more red maple trees (*Acer rubrum*) have been observed growing inside the stand of tamaracks that border a small pond. It looks as though the red maples will eventually shade the tamaracks and the other bog plants, thereby eliminating them.

Your Challenge: *Help find factors that have caused the changes at the Rose Lake marsh, and predict what we can expect to happen to the bog ecosystem over time.*

In this story, the boxed text gives a context for the story. The shaded text gives some information that may be relevant to the challenge. The underlined text states the challenge presented to students. The rest of the story gives details that are not essential to solving the problem.

Often, teachers naturally have the urge to write more details into the story, but keep in mind that Page 1 is intended to *start* the conversation, not answer all of the questions. One of the things teachers are tempted to do is to define terms in the story. In this case, students may not be familiar with the term *bog*, but the structure of the PBL analysis framework gives learners a chance to list "What is a bog?" as a "need to know" issue. Even if terms are introduced that you suspect students won't know, avoid adding a definition. In Page 1, you only need to spark some curiosity, and students will let you know when they need a definition or explanation.

Writing Page 2

Page 2 of the story provides a bit more information, as shown in the Bogged Down example (p. 134).

Page 2: More Information

Bogged Down

A bog is a standing body of water with no underground spring of freshwater to feed it. **The water is generally cold, extremely acidic, and low in oxygen. Sphagnum moss grows in the bog and forms a thick mat of floating plants.** *These plants, over time, can fill in the pond or small lake with peat that will eventually be firm enough to support trees.*

In the middle of the bog is an area of open water. Around that is a border of mint and cattails, and just a few feet closer to the center is a mat of sphagnum moss and other plants that is so thick in spots a person can walk on top of the mat without falling into the water. The whole mat moves up and down. As a result, these ecosystems are sometimes called "quaking bogs." **This peat bog stage is followed over time by shrubs and tamaracks.**

Bogs have very little decomposition of organic matter because sphagnum gives off hydrogen ions, creating a very acidic soil. In this nutrient-poor soil, some plants have adapted by becoming carnivorous. Examples include sundews and pitcher plants. These plants trap insects to supplement their photosynthetic diets.

This new page in the story adds some information to help the learners move closer to a solution to the problem, but it still leaves out some essential ideas. The italic text directly answers questions that are likely to arise during the Page 1 discussion, including "What is a bog?" and "Why are trees starting to appear in the bog?" But there will likely be other "need to know" items that Page 2 does not answer. It is important to avoid giving all of the information needed to solve the problem, because the research phase helps students develop important skills, too.

The bold sentences are new information that is important in understanding why bogs change over time. The other text (roman type) includes interesting bits of information that may help students understand bogs but might not be needed to help students reach the ultimate learning goal. Having both relevant and non-essential new information on Page 2 is important. Remember that identifying useful and extraneous information is a part of the process of solving real-world problems. Your students need to practice this skill.

If the two pages of the story are written well, your students will have enough hints to keep their focus on the relevant science concepts, but they will still need to think about new questions that will require further research. As you gain more experience facilitating PBL lessons, your sense of how much information to include or leave out will become keener, and the initial phases of writing a problem will become easier.

Getting Feedback

When you have a draft of a problem that seems appropriate, we suggest asking someone else to read it. If possible, have a nonscientist or someone who is not an expert in that particular science concept read the draft. This type of review lets you see from a new perspective whether the story is engaging and whether the information included is too much or too little. Ask your reviewer to tell you what science concept he or she would use to solve the problem. The answer may suggest that the story needs a bit more information to start your learners down the path you intended. If the reviewer can tell you the entire solution just from the Page 1 story, you may have provided too much information, assuming your reviewer is not already familiar with that particular story or context.

This review is an extremely valuable step! Once you have more experience writing problems, you might elect to skip this step, but the information you get from this review is so important that we suggest you always have another person read your story as you develop it.

Tips for Writing the Story

As you write, try to avoid writing the problem like a textbook question. The language you use should be comfortable and accessible to students and should sound more like a newspaper article or a conversational story. One strategy is to make the story a dialogue between characters. You may also be able to adapt "cases" from other sources (Rose, Schomaker, and Marsteller 2015) to fit within the PBL framework. You can use the problem stories in Chapters 5–8 as examples, and you will find several "styles" represented in those examples. And it is fine to have some fun with the story!

But the writer also needs to work to make the story believable. The ideal story is a true story. The story in the Where's Percho? problem in Chapter 6 is directly out of local news stories. In most cases, these are extremely easy stories to write. Others have been fictionalized, but they are real problems. An example is Calico Cats in Chapter 7: the characters in the story are fictional, but the questions are real, and the story could very well happen in the real world. Several other problems in this book are also fictionalized. The key is to not make up a story that is beyond the realm of possibility. The more realistic, the better! Box 9.1 (p. 232) gives more tips for writing a PBL story.

Integrating Investigations

Sometimes students will identify "need to know" issues that could be addressed with a hands-on, inquiry-oriented investigation, such as the dissolved oxygen tests used in the Bottom Dwellers problem in Chapter 6. As you plan your own PBL problems, you will likely face the question of whether to plan to include one of your existing lab activities as a part of the lesson plan. Investigations are certainly an essential part of the science curriculum, and they have a place in the PBL framework. However, you may need to modify how you present the investigation to students. In this section, we will suggest considerations you should make as you find a way to fit a lab activity into your PBL problem.

The first consideration is the fit of the activity with the learning objectives of the PBL lesson. Early in your planning, you identified the desired outcomes. Does the lab you plan to include in the plan directly address those outcomes? If not, a better place for the investigation might be before or after the PBL problem. If the lab does contribute to students' learning of the target concepts, then it might be an appropriate strategy.

For instance, you might have a very good lab activity in which students look at karyotypes (photos of chromosomes in a cell) that are used to diagnose trisomy or monosomy disorders such as Down syndrome or Turner syndrome. Would it fit during the genetics problems about coat color found in Chapter 7? Probably not! The central concepts of those labs are about sex-linked traits and codominance, not trisomy or monosomy. The karyotype lab is valuable, but it would be more appropriate after the genetics PBL problems have been completed.

If the activity you are considering does fit the goals of your lesson, you should also think about how the lab is presented to students. In many lab activities, the preliminary materials and the handouts with the lab will give much of the information students might need to answer the investigation. Many of these labs are described as *confirmation* labs (Bell,

Box 9.1. Tips for Writing a PBL Story

- Use language that is accessible to students.
- Make the story as realistic as possible, even if it is fictionalized.
- Page 2 should answer some but not all of the "need to know" issues from Page 1.
- Have a nonscientist review the story. Ask if they can tell you what problem the learners need to solve and a possible solution.
    - If they can solve the problem without other sources, you've given too much information.
    - If they cannot identify the problem or connect to some broad science topic, you have not provided enough information.
- Add some color to the story with information that is not directly connected with the story, but not too much.
- Keep each page of the story short.

Smetana, and Binns 2005). To fit within the framework of a PBL problem, the lab activity should provide some important piece of information that will fit within the entire collection of evidence students gather. A lab activity will not replace the research phase but will complement it.

So examine the lab handout. Is there a specific question presented as the focus of the investigation? Does it include follow-up information that gives away too much about your problem too soon? If so, revise the handout. You can remove some of the background information or replace it with your own version that fits within the context of the story you have written. You can also write a testable question that narrows the students' focus to the evidence you need them to gather for the PBL problem. Any follow-up questions should be centered on having students relate the data and evidence to the problem.

Taking time to ask questions about the fit of the lab to your PBL problem is very important. If the investigation is only tangentially related to the learning goals, the lab will be a distraction, and students may lose momentum in their drive to solve the problem. If the lab handouts overtly present the answer to your problem, the investigation negates the need to do further research, or it may remove the opportunity for students to construct their own solutions. All of these are critical elements in learning science and should not be skipped.

The following tips summarize things to consider when integrating an investigation or hands-on lab as part of a PBL lesson plan:

- Check to ensure the lab addresses your learning goals.
- Remove excess background and follow-up information.
- Write a testable question that identifies the purpose of the lab.
- Let the investigation focus on *one part* of the evidence needed to solve the problem.

Identifying Potential Resources for Students

When the story is written, one of the next steps is to look for sources of information that your students will need to find in the research phase of the PBL process. In this stage, students will search the internet, textbooks, library materials, and any sources you provide for information that will answer the "need to know" items they feel are most relevant to the problem.

The teacher has different options for providing those resources or access to the resources. Many facilitators let learners search the internet in class, on tablets or laptops, or in a computer lab. There are many search engines that you can use to find sources of information for most authentic problems.

However, students may find that there are so many resources that they are overwhelmed by the sheer volume of content they might need to sort through. If your students are not ready to discuss strategies for narrowing their search terms or evaluating the strengths and weaknesses of different sources, you may find it helpful to locate several sources and provide a set of links for students to use (see Box 9.2). If your school has a course management system or a course website that students can access, you can post links for students to choose. Experience tells us that you can provide more links than learners need, and your students can begin to practice strategies for selecting sources from a list.

In some classrooms, or with some students, especially younger learners, teachers may not be comfortable sending the class to the internet to do a search. In these cases, you can make copies of some sources from websites, newspapers, magazines, books, and other texts. Make sure you comply with copyright and fair use practices with this option. With a set of resources in hand, you can produce packets of related information.

If you provide the research materials, there are a couple of useful strategies for sharing the information with student groups. One is to provide a packet with all the materials for each group and allow them to select the files that are more important for their research. This is an especially effective approach if you have asked each group to develop its own solution and share its final answers with the class. Each group will receive all the text materials, so they will have the necessary information to build a solution.

For a lesson in which each group selects different "need to know" issues to research, the teacher can sort files by topic and give each group a different packet. The advantage of this strategy is that the groups will need to share their findings with each other to enable the entire class to construct a solution. This process models the work of real scientists and helps build a sense of teamwork in the classroom.

Whichever strategy you decide to use for the research phase of your lesson, you need to make sure there are relevant sources available to students. Take the time to do searches using terms you think students are likely to use, and check the sources to make sure they are reliable and scientifically accurate. One way to evaluate the trustworthiness of the source is to check the author's credentials. If he or she is a qualified authority, the source is more likely to be accurate. But remember that just about anyone can post a website as a "source," so not all the content you find will list an "expert" as the author.

You may also want to check the URL. Web addresses that end in *.edu* and *.gov* are usually more reliable, although many *.com*, *.org*, and *.net* sources also contain excellent information. Sites posted by nonprofit organizations such as nature centers, conservation groups, and citizen science projects will have a *.org* suffix, but they can be some of the best sources.

When you do this search, you should use a computer in your classroom or school. If you find a source at home, you may find that the students cannot access the site if your school has a proxy server or a filter that blocks some websites. By locating sources in the same location where your students will conduct their searches, you can make sure the resources will work for your class when they need them. Identifying relevant sources in advance will also let you submit a request to your network administrators to give permission in the filter for the sources you need for a given lesson.

Printing copies of the sources you find may also be a good idea, even if you plan to let students do their internet searches. Technology can have a glitch at any moment, and if the server hosting the site you want students to visit is not working, or if the organization moves the site or redesigns their page, your students might run into broken links or search lists that lead to a different page. Be prepared for the possibility of a technology glitch!

Box 9.2. Tips for Locating Sources

- Use multiple search terms that students may select.
- Check sources to see if they are reliable and accurate.
- Print backup copies in case links are not working.
- Search from the classroom to see what your school's filter will allow.
- Submit in advance a filter exception for any sites students need to use.
- Keep a list of links available on a course website in case you need it.

Writing the Solution

Just as when you write your assessment questions, one important step is writing a model response for the solution to the problem. It can be easy to overlook this step as you plan your PBL lesson. Having a solution written before presenting the problem to students will help you plan to evaluate the final solutions your students produce.

But your solution needs to be flexible, or you may even need to include more than one possible answer. Some PBL problems may have more than one acceptable solution. In most of these cases, students may have different sets of values they use when selecting a solution. For instance, any problem relating to a plan to use natural resources may lead to different choices. Students may choose to preserve a habitat or propose a responsible strategy for using a resource such as lumber or minerals; the choice depends on the degree to which students value conservation of the natural site versus balancing human and economic resources. Be prepared for the possibility of multiple solutions.

Writing a model solution can also help you ensure that the wording of the story, the available resources, and the challenge presented to students are designed to permit learners to reach an appropriate solution. You may find that you need to revise the story or locate additional materials to foster your students' successful completion of the lesson.

Piloting a Problem

Another strategy to try as you plan a PBL lesson is to pilot the problem with a small group of volunteers. Try to create a group that is not familiar with the content and ask them to read the story. By letting them apply the PBL analysis framework (What do we know? What do we need to know? Hypotheses?), you can find potential sources of problems your students would encounter. As with having a reviewer read the story, a pilot group can be a great source of feedback to catch issues you might overlook.

Some possible groups from which you can build a pilot group include a science club or other student group, family members, or fellow teachers in other subjects. If your school has created professional learning communities (PLCs) for school improvement, your PLC group may be an ideal place to share your lesson ideas, including a request to pilot the problem.

The pilot group will be better able to provide useful feedback if you give them questions to consider as they review the problem. The PBL Lesson Talk-Through at the end of this chapter (also available on the book's online Extras page at *www.nsta.org/pbl-lifescience*) can serve as a feedback guide for the pilot group participants.

Modifying Existing Lessons

If you are like most teachers, you probably have a very full curriculum plan, and adding new lessons to the schedule will only work if you drop another lesson. Moreover, if you are like most teachers, you probably also have lessons or activities in your files you really like, but you'd like to update or modify. PBL may be the format that will let you revise existing lessons. Modifying what you have is a great way to keep the ideas that work while still implementing new strategies.

So think about your current set of lesson plans. Do you have a lab activity, assessment question, or discussion starter that you feel could be more engaging? Does it relate to a real-world problem that your students can relate to? If so, these are great opportunities to modify what you have! Let's look at an example of a way to adapt your current activities.

In many biology classes, teachers include a lab about burning peanuts or almonds to measure the calories stored in foods. One aspect of the lab is to weigh the nut before and after you burn it to find out how the mass changes. The information from that is important in helping students recognize that matter is being changed in a chemical reaction that releases energy. In living cells, that reaction is cellular respiration. This lab is a good activity and has many practical applications. But one way to help students recognize similar chemical changes as energy conversions is to use the same concept in a different example.

Chapter 8 includes a problem called Mysterious Mass that is based on the concept of respiration as an energy conversion.

For Mysterious Mass, the process of modifying the existing lesson began with brainstorming some potential contexts that can demonstrate the change in mass when organic compounds are burned to release energy. Although there are several possible examples, the authors of this problem chose the campfire example to fit within the storyline of the other problems in this chapter.

The authors then drafted a story that sets the scene that students will think about as they explore the energy conversions of cell metabolism. They shared the story with colleagues to get feedback, made some revisions, and wrote some new assessment questions. The lesson still leaves an opportunity for an inquiry investigation with the peanuts or almonds using the existing lab activity, but the topic is now presented as a PBL problem.

Your own curriculum plan is a potential source for PBL ideas. Take the time to check your current set of lessons and see if you can find a place to be creative in writing a new PBL problem!

While we are thinking about modifying lessons, please remember that problems are often specific to a particular place and time! Chapters 5–8 included lessons that were written for learners in Michigan. Some of the local issues that led to story ideas may not be as engaging for your classroom. So feel free to change the stories if you find it helpful! Modify what we have presented, just as you may choose to modify your own lessons.

Resources for Writing PBL Problems

At the end of this chapter are two aids that you can use as you plan your own lessons. The first is a PBL Lesson Template (p. 239) with prompts to help you think about the parts of the PBL lesson structure we have presented. The second is a PBL Lesson Talk-Through (p. 240), which includes a list of questions you should think about and discuss as you plan and revise your lesson. Both of these documents are also available on the book's online Extras page at *www.nsta.org/pbl-lifescience.*

References

Bell, R. L., L. Smetana, and I. Binns. 2005. Simplifying inquiry instruction. *The Science Teacher* 72 (7): 30–33.

Boersma, K., L. Ballor, and J. Graeber. 2008. Poison ivy: Teachers designing a problem-based science unit. *MSTA Journal* 53 (1): 28–31.

Bransford, J. D., and D. L. Schwartz. 1999. Rethinking transfer: A simple proposal with multiple implications. *Review of Research in Education* 24 (1): 61–100.

Rose, J., B. Schomaker, and P. Marsteller. 2015. CASES Online. Emory University. *www.cse.emory.edu/cases.*

Wiggins, G. P., and J. McTighe. 2005. *Understanding by design.* Alexandria, VA: Association for Supervision and Curriculum Development.

PBL Lesson Template

Authors:

Grade Level:

Science Content Topic:

Expected Timeline:

"Big Idea":

Standards:

Story—Page 1: *(narrative giving context, just enough to identify the challenge)*

Story—Page 2: *(additional necessary information, more details on the story)*

Potential Sources:

<u>Assessment</u>

Pre/Post Assessments:

Transfer Task:

Solution/Presentations:

Informal Assessments:

PBL Lesson Talk-Through

Name of Lesson:

Talk-Through From Student Perspective

Pretend to be one of your students. Without thinking about the answer or solution, answer the following questions:

- How would you interpret the task/challenge? What is the problem asking you to do?
- What questions come to mind that might help you solve the problem?
- How would you go about solving this problem?
- What would you as a student know that would be useful in solving the problem?
- What similar tasks might you have experienced before?
- What is confusing about the problem?

Talk-Through From Teacher Perspective

Answer the following questions as the teacher/facilitator for the lesson:

- What would you expect your best students to do with this problem?
- What would you as a teacher expect your less able students to do with this problem?
- What materials will be required? What resources?
- Which resources might be difficult for your students to find?
- How long do you estimate this lesson will take?

IMAGE CREDITS

Chapter 1

Figure 1.1: NSTA Press

Chapter 3

Figure 3.1: Janet Eberhardt

Chapter 4

Figure 4.1: Tom J. McConnell

Figure 4.2: Tom J. McConnell

Chapter 5

Figure 5.1: Photo0pal, Attribution license. *www.everystockphoto.com/photo.php?imageId=3565977&searchId=6cd38319dbd99aab5499918ef430ebc5&npos=1.*

Figure 5.2: Ricardo Rodríguez, Freeimages.com Content license. *www.freeimages.com/photo/kruskal-1-1532012.*

Figure 5.3: Crissy Pauley, Freeimages.com Content license. *www.freeimages.com/photo/grandma-s-love-1510211.*

Figure 5.4: just4u2009, Attribution license. *www.everystockphoto.com/photo.php?imageId=10465536&searchId=f25df9614f6c31844cf901a8c6cfd7dc&npos=68.*

Figure 5.5: platycryptus, Attribution license. *www.everystockphoto.com/photo.php?imageId=25218934&searchId=3e57895dab3cb5642ab1d15ea97135c&npos=113.*

Figure 5.6: NaturalMary63, Attribution license. *www.everystockphoto.com/photo.php?imageId=6060217&searchId=17076a36461b459d15e49e471f73aecb&npos=21.*

Figure 5.7: eef ink, CC BY-NC-ND 2.0. *www.flickr.com/photos/eef-ink/6194994134/in/photolist-arqXVG-3DQ83-9k5L4K-9L8poB-amoX6X-iA195-5fTuSp-gj774M-6pXrGj-urtu5w-2Xoh1-PBAa1-povBFf-aeWVXc-ftGc9E-bVjhfj-krcgq-cBitRA-7M3GWQ-8Gfc6N-nctssB-4cR6r-8yiWBA-89vK2W-agjfwu-98S1cV-epg8h1-bnErLb-2vup9K-ay1ZpH-oa21dK-9LCPv8-7fhNvk-cSVUqL-9sbKqx-6w8NNw-5o2EqC-88vN37-mStPGn-3aqjD1-3D894g-6PteXU-7uxin9-bGNBRe-fq2txU-9Ti5SU-th1BvR-rmQ4xg-6VBWA7-4KaZhN.*

Figure 5.8: Peter Stevens, CC BY 2.0. *https://commons.wikimedia.org/wiki/File:Golden_Ground_Mushrooms_(15529795166).jpg.*

Figure 5.9: TheAlphaWolf, GNU Free Documentation License. *www.everystockphoto.com/photo.php?imageId=1611340&searchId=4ad3b50b13f1a572d3e3c669e0190b15&npos=5.*

Figure 5.10: Clearly Ambiguous, Attribution license. *www.everystockphoto.com/photo.php?imageId=583093&searchId=54e12f98c3fc7fda92c0b05c7c42e82f&npos=30.*

Figure 5.11: ndrwfgg, Attribution license. *www.everystockphoto.com/photo.php?imageId=878315&searchId=b7e7c3ce787412ee01954ab0f4c82d0f&npos=.*

Figure 5.12: m.derepentigny, Attribution license. *www.everystockphoto.com/photo.php?imageId=4450672&searchId=48e664afb19ab4a186061f4851caa700&npos=171.*

Figure 5.13: skyseeker, Attribution license. *www.everystockphoto.com/photo.php?imageId=3250371&searchId=2df6def39a07de8246a8e23b067804d9&npos=185.*

Chapter 6

Figure 6.1: magurka, Freeimages.com Content license. *www.freeimages.com/photo/cormorant-1-1249825.*

Figure 6.2: Otis Maha, USFWS, Attribution–ShareAlike license. *www.everystockphoto.com/photo.php?imageId=22054&searchId=b4421611ed5efd45ea772c77e26a002d&npos=17.*

Figure 6.3: mike baird, Attribution license. *www.everystockphoto.com/photo.php?imageId=2605901&searchId=b4421611ed5efd45ea772c77e26a002d&npos=125.*

Figure 6.4: Ingrid Taylar, CC BY 2.0 (branded sea lion) *www.flickr.com/photos/taylar/9725507693*; Benjamin Earwicker, Freeimages.com Content license (sea lions swimming) *www.freeimages.com/photo/spare-parts-2-1353949*; Eric Guinther, GNU Free Documentation license (fish bypass) *https://commons.wikimedia.org/wiki/File:BonnevilleDam.jpg.*

Figure 6.5: NSTA Press

Figure 6.6: echoforsberg, CC BY 2.5. *www.everystockphoto.com/photo.php?imageId=10551253&searchId=d87ac001ae36445f3cb8de2df9b48b09&npos=56.*

Figure 6.7: NOAA Restoration Center, Jim Turek, Public Domain. *www.photolib.noaa.gov/htmls/r0006642.htm.*

Figure 6.8: Dave Huth, CC BY 2.0. *www.flickr.com/photos/davemedia/5789175336/in/photolist-9Pz38S-ia4xzy-f5Wj3K-f6byUG-f5pWg8-fdH5g2-etyS5e-6sTQyu-6BK4nE-f5WiW4-dqu1AM-hAfeUX-fdXnRW-f5Ebvo-4UgzmH-fkwFuo-4UgzVM-ed5ntx-oFkBC6-oFoFyu-dmYYgi-cpQ8Wj-ciNnAw-a2Yfe8-bCcY8x-rNgTvr.*

Figure 6.9: weatherbox, Freeimages.com Content license. *www.freeimages.com/photo/mayfly-1551709.*

Figure 6.10: Tobyotter, CC BY 2.0. *www.everystockphoto.com/photo.php?imageId=21122316&searchId=00793624942d13e12b178513d2058a54&npos=235.*

Figure 6.11: USFWS Pacific, CC BY 2.0. *www.everystockphoto.com/photo.php?imageId=14573096&searchId=c8384e7f145836d182bd6b27c67952b3&npos=149.*

Figure 6.12: Steven Katovich, USDA Forest Service, Public domain. *www.forestryimages.org/browse/detail.cfm?imgnum=5492753.*

Figure 6.13: Denali National Park and Preserve, CC BY 2.0. *www.flickr.com/photos/denalinps/5302689686/in/photolist-95zEUf-7oyw92-7oCpjQ-rH2nVw-EyLus-26Bz9-7Zzsg-7Gwf1X-pVTacK-dJsMG4-9JjMjK-8oqfDA-57pjNw-trXJvo-tJEEsi-3JEauy-4EQY54-pVvjFF-b5EXM6-8Fzp7k-3JzPGZ-4hUPBL-gqDfPs-eGNyo-5F7KFr-a7PHBc-fmrwVs-ppMKya-6ys3gW-7v8MnM-6r6V4H-5gHyeX-PX7Y7-buwork-c5PcXs-8J1kWQ-a7SzFW-5Q1k56-5YMT8j-f4HSQx-62opYD-oDERvh-gSN5ni-3Wjnb8-66PyJz-cFKUW1-96XZFN-cAr5AU-8icZDz-h3qHvH.*

Figure 6.15: NSTA Press

Figure 6.16: Tom J. McConnell

Figure 6.17: Wilson 44691, CC0 1.0 Universal Public Domain Dedication. *https://commons.wikimedia.org/wiki/File:Sphagnum_Brown%27s_Lake_Bog.jpg.*

Figure 6.18: Nicholas A. Tonelli, CC BY 2.0. *www.flickr.com/photos/nicholas_t/9340814208.*

Figure 6.19: Muffet, Attribution license. *www.everystockphoto.com/photo.php?imageId=7504923&searchId=c410a282a5b3ffcab32f21cd59954b9d&npos=9.*

Figure 6.20: Liz West, CC BY 2.0. *www.flickr.com/photos/calliope/65205564/in/photolist-6LckS-gxhME-gxk6d-7WLWm5-c1oAKj-aXpnHK-mk6oAb-6gxyyy-7QPzqf-ai7NmJ-71vUNH-bZsnoJ-cZojMU-apxjKw-6e33Kt-aibYwm-7wZc9m-aKLrAD-bFfaKb-98NBkL-4b2CMG-pehB1B-dapjAz-mM8qu-4JTuyS-7yxZ2g-rsPT2d-gGgMK2-Hc1HZ-5e6Bbz-a7AfWo-aDtoiB-98NBsb-4zW4vG-qFFfuE-2zjkPh-efFUeM-8XzDPA-7LSQQ9-76qsiW-6Lcdn-dnNW4J-yfcNZ3-nEH45T-84a4UN-wMVrtm-gxi5i-aiwMfv-5jqj5s-962qry.*

Figure 6.21: Joyce Parker

Figure 6.22: NSTA Press

Chapter 7

Figure 7.1: NSTA Press

Figure 7.2: Laitche, Public Domain. *https://en.wikipedia.org/wiki/Kitten#/media/File:Laitche-P013.jpg.*

Figure 7.3: Jon Ross, CC BY 2.0. *www.flickr.com/photos/jon_a_ross/2339928055.*

Figure 7.4: Tom J. McConnell

Figure 7.5: NSTA Press

Figure 7.6: brokinhrt2, Attribution license. *www.everystockphoto.com/photo.php?imageId=6001361&searchId=d5cc925026174a15a77191000aa2e641&npos=213.*

Figure 7.7: Eirik Newth, CC BY 2.0 *www.flickr.com/photos/29904699@N00/286759956* (gray and white cat); Melissa Roadruck, with permission (black cat).

Figure 7.8: (from left to right) BFS Man, CC BY 2.5, *www.flickr.com/photos/bfs_man/5382155817*; Creating Character, CC BY 2.5, *www.everystockphoto.com/photo.php?imageId=11239809&searchId=f7ca2a95309079d4b0d850cab3b6ce59&npos=18*; Sandy Schultz, CC BY 2.0, *www.flickr.com/photos/chatblanc1/4969653076.*

Figure 7.9: (from left to right) mconnors, MorgueFile licens. *http://morguefile.com/archive/display/5388*; Alvimann, MorgueFile license. *www.morguefile.com/archive/display/226430.*

Figure 7.10: NSTA Press

Figure 7.11: Mikee032901, MorgueFile license *http://morguefile.com/archive/display/100329* (top left); Monica McConnell, with permission (top right); packrat, MorgueFile license *http://cdn.morguefile.com/imageData/public/files/p/packrat/hr/fldr_2005_05_17/file000348639262.jpg* (bottom left); aturkus, Attribution license, *www.everystockphoto.com/photo.php?imageId=2684957&searchId=0832c1202da8d382318e329a7c133ea0&npos=1304* (bottom right).

Figure 7.12: Tom J. McConnell

Figure 7.13: NSTA Press

Figure 7.14: Laitche, Public Domain. *https://en.wikipedia.org/wiki/Kitten#/media/File:Laitche-P013.jpg.*

Figure 7.15: Lil Shepherd, Attribution license. *www.everystockphoto.com/photo.php?imageId=17456859&searchId=06ae4e03b831d66859cf8613bb96e52a&npos=79.*

Chapter 8

Figure 8.1: Kirt Edblom, CC BY-SA 2.0. *www.flickr.com/photos/27190564@N02/18642860771/in/photolist-84Uipp-9Pqccc-kA2wV-4eBeM6-6Cg9ya-ner9g-8dzjxN-8dzuR7-8dze5f-71RtjV-8dtEZp-8dw8ar-hWaSJb-8dzmv1-8dwUAA-8dwfPp-6Cg8wg-6DFc3u-4aFp3e-8dhXxg-6DB3wi-nmWzV-5fB5a4-3PbiV-4eBeQ8-4kgYFQ-6DB3bi-74CFD-7hz82y-6Cg74c-4eBeSX-5mzhuJ-6Ckdb7-6Cg85H-yA6fmv-6CkdSh-6Ckfpd-6Cg7MF-ktL2L-4EU5yh-sUZY4-uppwZZ-uk3ukD-u7KGTB-toQVmi-u47gj3-toESjY-sRgoWn-x8qian-wdhtgP.*

Figure 8.2: Joyce Parker

Figure 8.3: Joyce Parker

Figure 8.4: Joyce Parker

Figure 8.5: Joyce Parker

Figure 8.6: MSVG, Attribution license. *www.everystockphoto.com/photo.php?imageId=11941819&searchId=e24204c6c885dbb1fc6a7fc76ee488de&npos=14.*

Figure 8.7: drbloodmoney, CC BY-SA 2.0. *www.flickr.com/photos/drbloodmoney/108793743.*

Figure 8.8: Joyce Parker

Figure 8.9: Scripps Institute of Oceanography, Public domain. *https://scripps.ucsd.edu/programs/keelingcurve/wp-content/plugins/sio-bluemoon/graphs/mlo_full_record.png.*

Figure 8.10: Kim, CC BY-SA 2.0. *www.flickr.com/photos/thegirlsny/6186177821/in/photolist-aqDM9i-7zkg8P-6ayHqD-aUTaL4-7jcg1u-9rj8ho-5oQ1B2-aFrC44-7vnZdk-6Z59TP-ig8syz-6cQtfQ-7uh4cr-agCxdc-99Sbp3-dL3F2M-7fYwzw-pNYXtr-cfjezo-5F-e1Cj-5QHM3n-6bq8Cj-fqTWVg-9XXdZ7-C4dqC-bxVdeD-koonaT-9XUk3P-8tDLD9-gh3ZNF-oap6bS-81qo3L-7ixxYz-d7v1Mo-biEeQc-adRCUJ-bpsyJf-93cBEL-7iSP8q-fqTMLk-jp59h-a77pzx-adNFA2-6rEqjA-n4NoHk-e9z4yD-75MWVF-95kF2s-dVWJLz-h7TFrn.*

Figure 8.11: takomabibelot, Attribution license. *www.everystockphoto.com/photo.php?imageId=1103561&searchId=54e12f98c3fc7fda92c0b05c7c42e82f&npos=12.*

INDEX

*Page numbers in **boldface** type refer to tables or figures.*

INDEX

INDEX

INDEX

INDEX

INDEX

INDEX